HIGHER

GEOGRAPHY
2007-2011

2007 EXAM – page 3
Physical and Human Environments
Environmental Interactions

2008 EXAM – page 39
Physical and Human Environments
Environmental Interactions

2009 EXAM – page 67
Physical and Human Environments
Environmental Interactions

2010 EXAM – page 95
Physical and Human Environments
Environmental Interactions

2011 EXAM – page 125
Physical and Human Environments
Environmental Interactions

ANSWER SECTION – page 157

SQA   BrightRED PUBLISHING

Publisher's Note

We are delighted to bring you the 2011 Past Papers and you will see that we have changed the format from previous editions. As part of our environmental awareness strategy, we have attempted to make these new editions as sustainable as possible.
To do this, we have printed on white paper and bound the answer sections into the book. This not only allows us to use significantly less paper but we are also, for the first time, able to source all the materials from sustainable sources.

We hope you like the new editions and by purchasing this product, you are not only supporting an independent Scottish publishing company but you are also, in the International Year of Forests, not contributing to the destruction of the world's forests.

Thank you for your support and please see the following websites for more information to support the above statement –

www.fsc-uk.org

www.loveforests.com

© Scottish Qualifications Authority
All rights reserved. Copying prohibited. No part of this publication may be reproduced, stored in a retrieval system, or transmitted in any form or by any means, electronic, mechanical, photocopying, recording or otherwise.

First exam published in 2007.
Published by Bright Red Publishing Ltd, 6 Stafford Street, Edinburgh EH3 7AU
tel: 0131 220 5804 fax: 0131 220 6710 info@brightredpublishing.co.uk www.brightredpublishing.co.uk

ISBN 978-1-84948-215-8

A CIP Catalogue record for this book is available from the British Library.

Bright Red Publishing is grateful to the copyright holders, as credited on the final page of the Question Section, for permission to use their material. Every effort has been made to trace the copyright holders and to obtain their permission for the use of copyright material. Bright Red Publishing will be happy to receive information allowing us to rectify any error or omission in future editions.

HIGHER

2007

[BLANK PAGE]

OFFICIAL SQA PAST PAPERS   5   HIGHER GEOGRAPHY 2007

X208/301

NATIONAL
QUALIFICATIONS
2007

MONDAY, 28 MAY
9.00 AM – 10.30 AM

GEOGRAPHY
HIGHER
Paper 1
Physical and
Human Environments

Six questions should be attempted, namely:

all four questions in **Section A** (Questions 1, 2, 3 and 4);

one question from **Section B** (Question 5 **or** Question 6);

one question from **Section C** (Question 7 **or** Question 8).

Write the numbers of the **six** questions you have attempted in the marks grid on the back cover of your answer booklet.

The value attached to each question is shown in the margin.

Credit will be given for appropriate maps and diagrams, and for reference to named examples.

Questions should be answered in sentences.

Note The reference maps and diagrams in this paper have been printed in black only: no other colours have been used.

Scottish Qualifications Authority

HIGHER GEOGRAPHY 2007

OFFICIAL SQA PAST PAPERS

1:50 000 Scale
Landranger Series

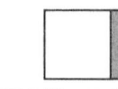

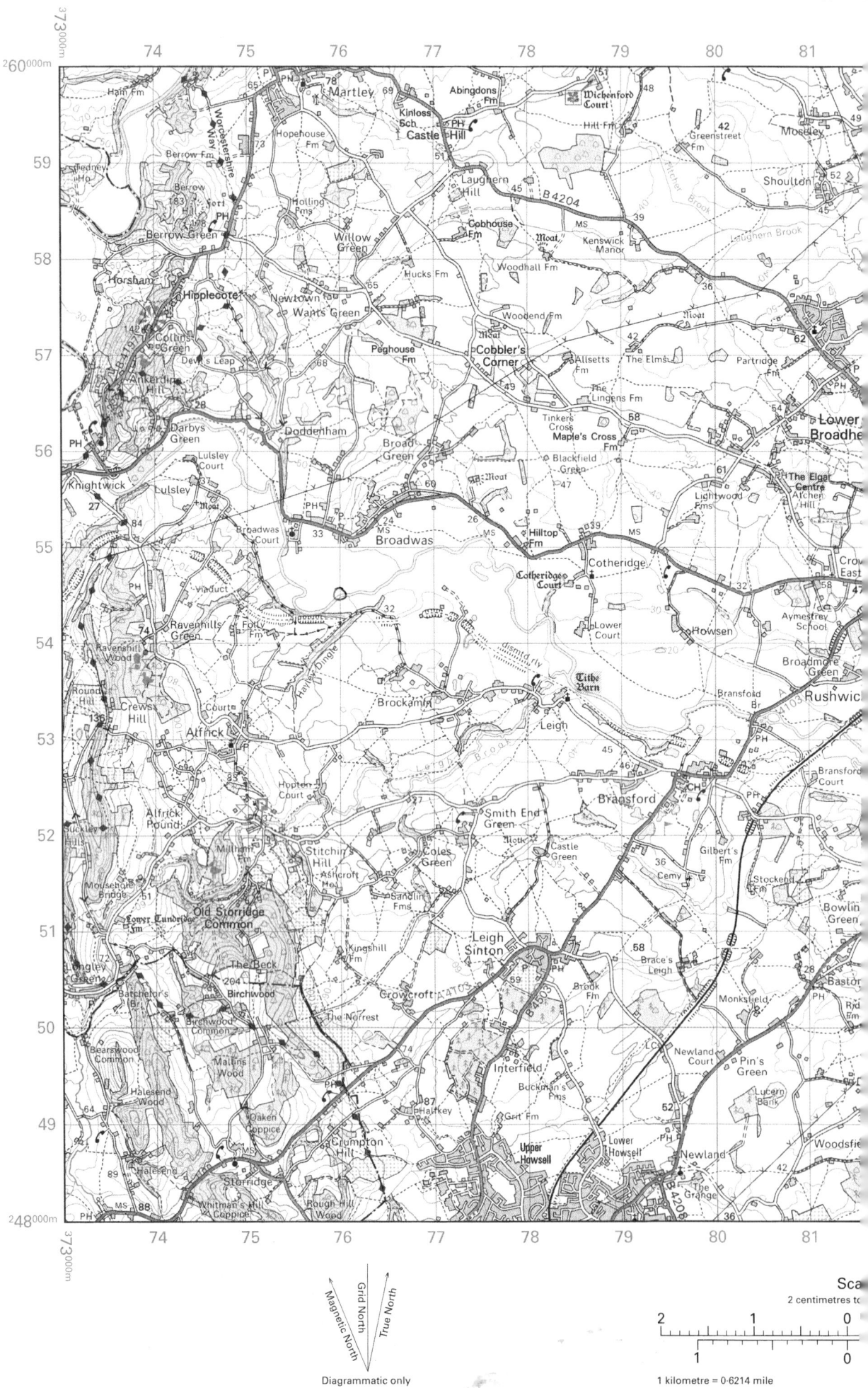

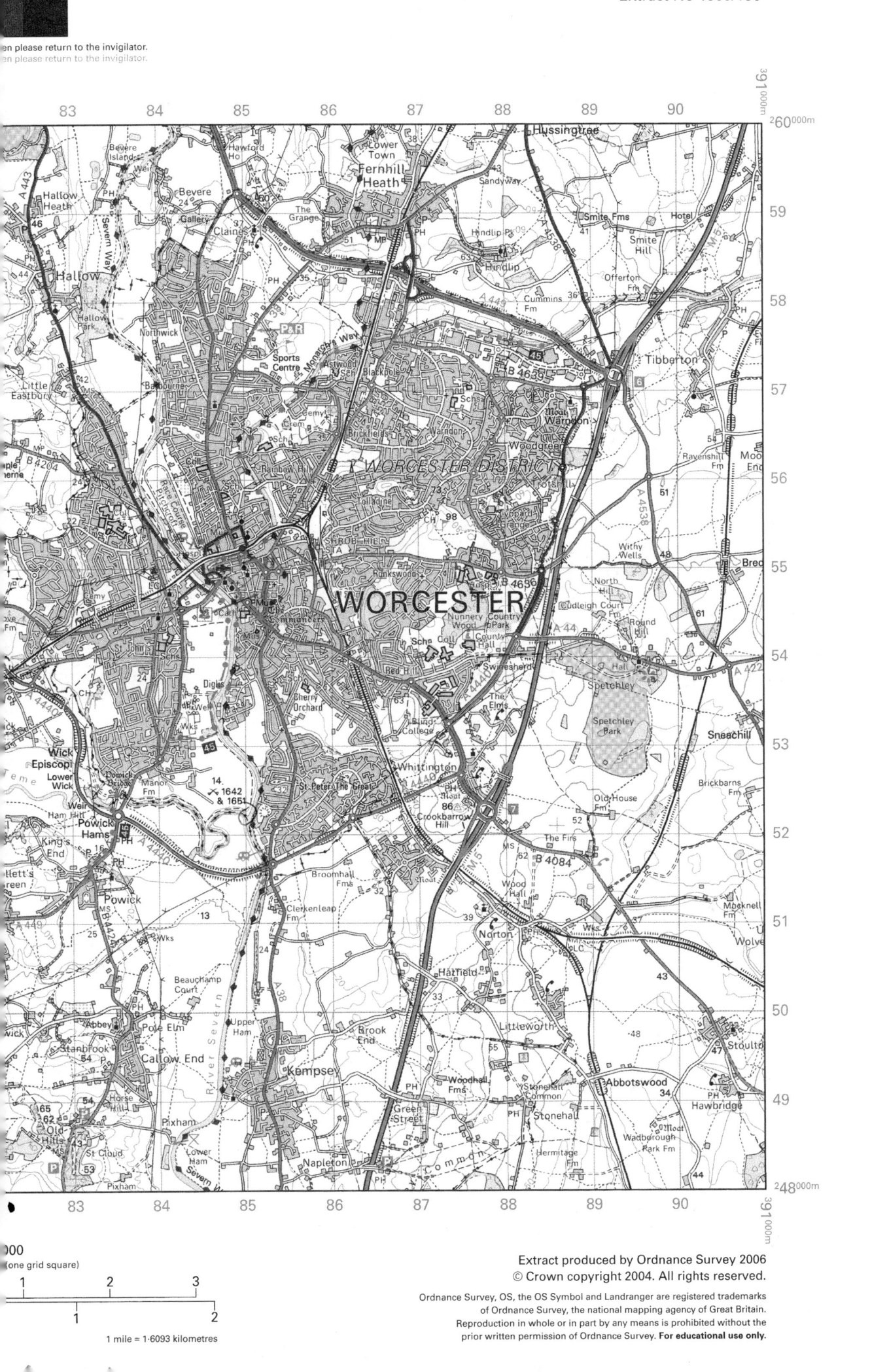

OFFICIAL SQA PAST PAPERS   HIGHER GEOGRAPHY 2007

Extract No 1560/150

Extract produced by Ordnance Survey 2006
© Crown copyright 2004. All rights reserved.

Ordnance Survey, OS, the OS Symbol and Landranger are registered trademarks of Ordnance Survey, the national mapping agency of Great Britain. Reproduction in whole or in part by any means is prohibited without the prior written permission of Ordnance Survey. **For educational use only.**

1 mile = 1·6093 kilometres

Marks

SECTION A: Answer ALL questions in this section

Question 1: Atmosphere

(a) Study Reference Diagram Q1A and Reference Map Q1A.

Identify air masses A and B, and **describe** their origin and nature.   3

Reference Diagram Q1A (The Inter-tropical Convergence Zone (ITCZ))

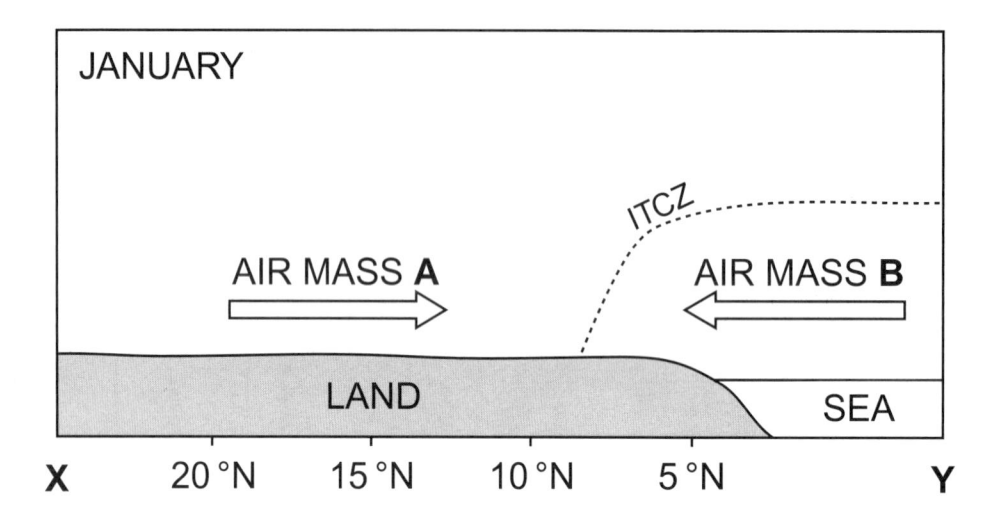

Reference Map Q1A (Location of section X–Y)

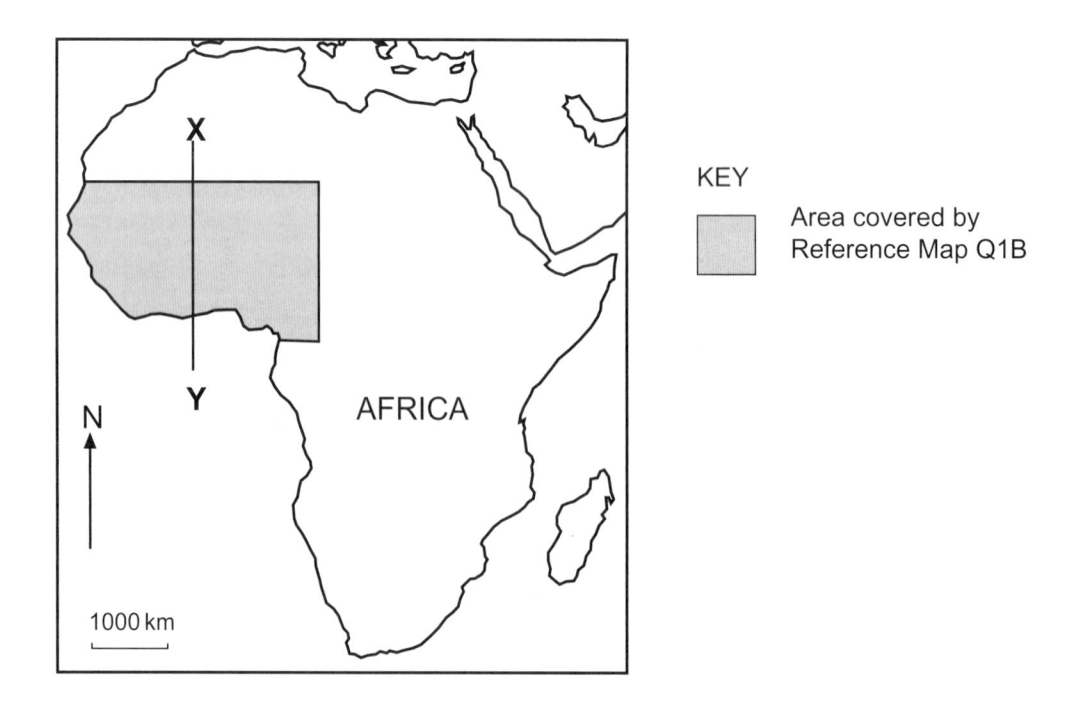

Marks

Question 1 (continued)

(b) Study Reference Map Q1B and Reference Diagram Q1B.

Describe and **explain** the varying rainfall patterns shown in Reference Diagram Q1B.

6

Reference Map Q1B (Rainfall patterns in West Africa)

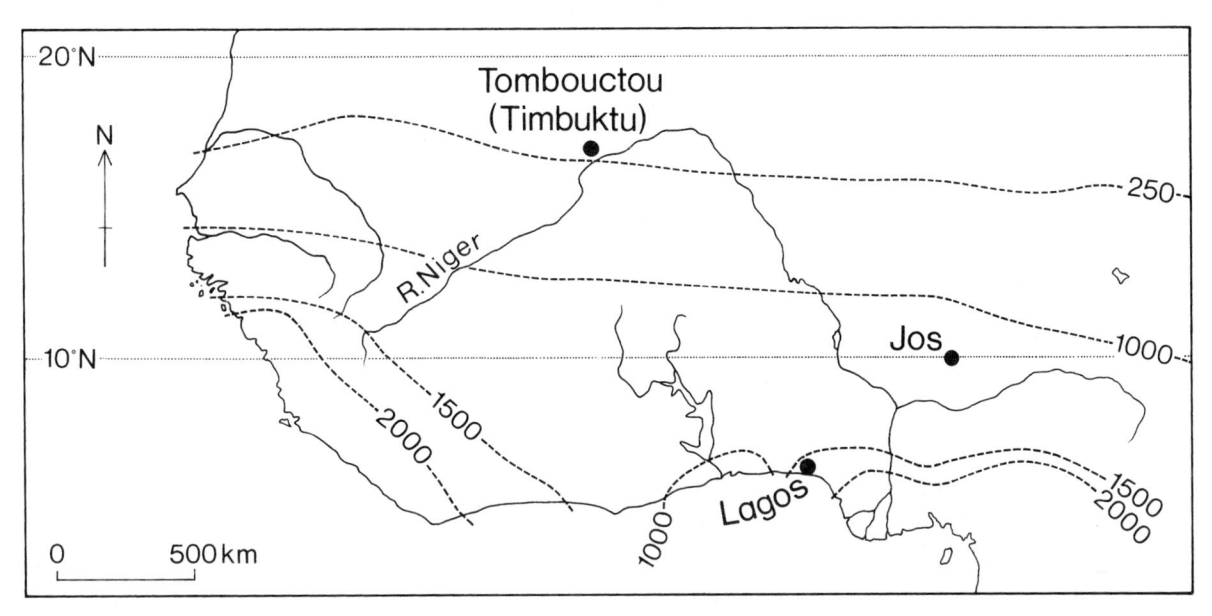

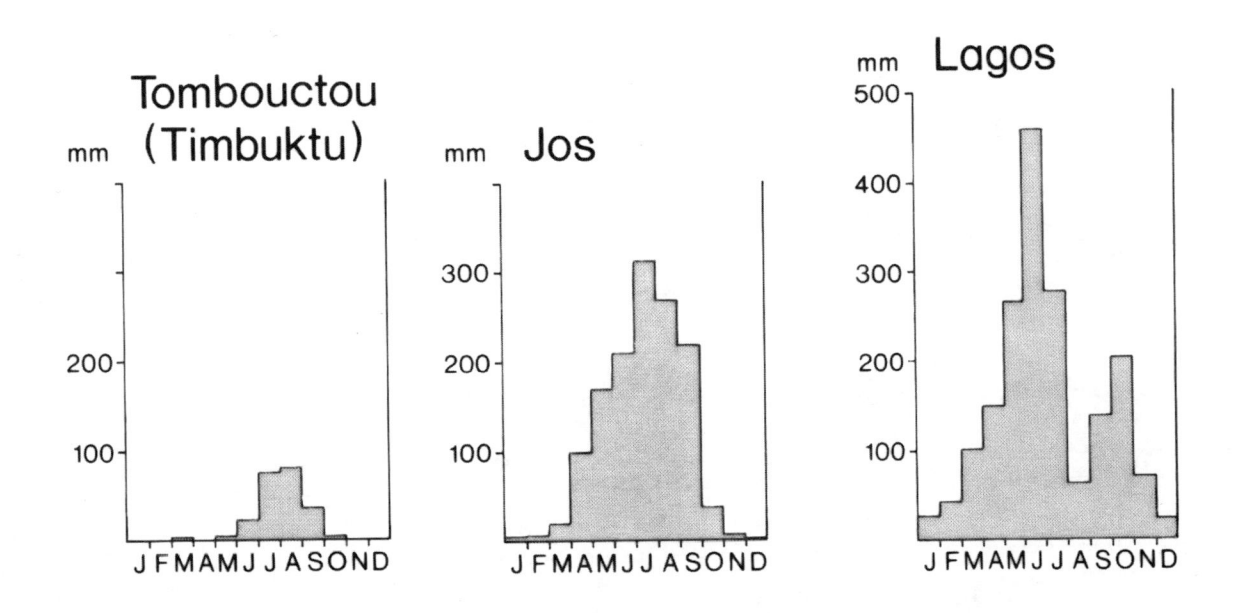

KEY ------- Isohyets showing mean annual rainfall (mm)
~~~~~~ Rivers

Reference Diagram Q1B (West Africa—selected rainfall graphs)

[Turn over

[X208/301] Page three

Marks

Question 2: Hydrosphere

Study OS Map Extract number 1560/150: Worcester (*separate item*).

(a) The River Teme is in the lower section of its course between 760545 and 850522. **Describe** the physical characteristics of the river and its valley in this section.   5

(b) Select **one** of the features that you have described for part (a) and **explain**, with the aid of a diagram or diagrams, how this feature is formed.   4

Marks

Question 3: Population Geography

(a) **Describe** the ways in which countries can obtain accurate population data. **3**

(b) **Explain**

(i) why **ELDCs** (Economically Less Developed Countries) may find the collection of such data more difficult, and

(ii) why the quality of data obtained may be less reliable than that gathered in an **EMDC** (Economically More Developed Country). **6**

[Turn over

Marks

Question 4: Industrial Geography

(a) For South Wales, or any other industrial concentration in the European Union which you have studied, **describe** the physical and human factors which led to the growth of traditional industries before 1950.   4

(b) Study Reference Map Q4.

Describe and **explain** the methods used to attract newer industries and investments to South Wales, or to any other industrial concentration in the European Union which you have studied.   5

Reference Map Q4 (South Wales: Location of industrial developments since 1950)

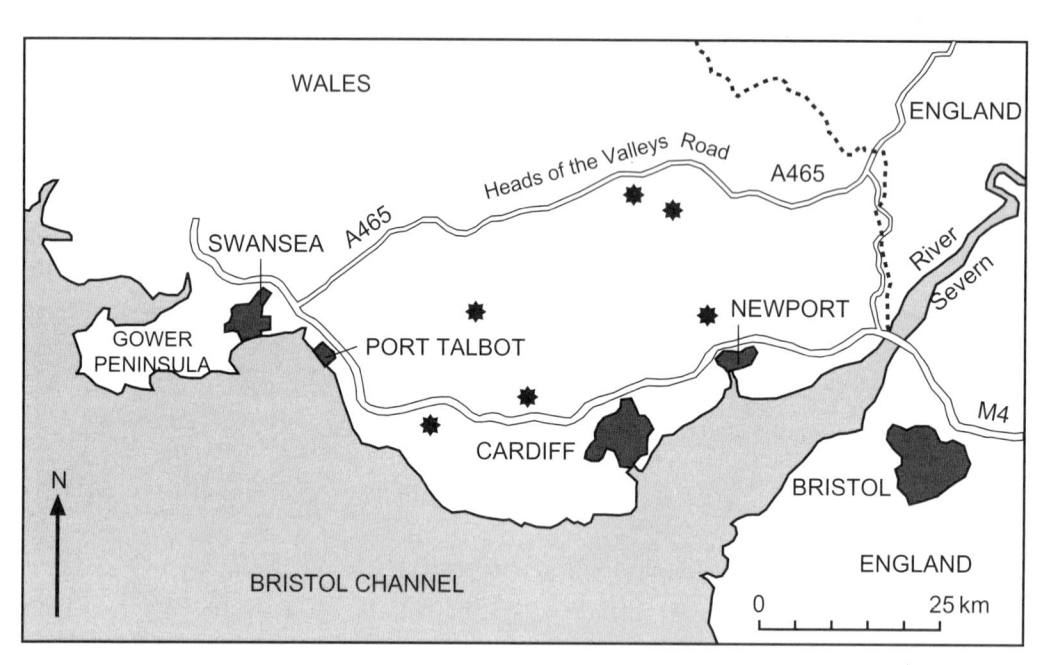

✽ Location of industrial developments since 1950

Marks

SECTION B: Answer ONE question from this section,
ie either Question 5 or Question 6.

Question 5: Lithosphere

Study Reference Photograph Q5 which shows a glaciated upland landscape in the Cairngorm Mountains.

(*a*) **Describe** the evidence which suggests that the area shown in the photograph has been affected by the processes of glacial erosion.   3

(*b*) Choose **one** feature of glacial erosion visible in the photograph and, with the aid of an annotated diagram (or diagrams), **explain** how it was formed.   4

Reference Photograph Q5

[Turn over

DO NOT ANSWER THIS QUESTION IF YOU HAVE ALREADY ANSWERED QUESTION 5

Question 6: Biosphere

Study Reference Diagram Q6A which shows two soil profiles.

Choose **one** of the soil profiles.

(i) **Describe** the characteristics of the soil, including horizons, colour, texture and drainage. **3**

Reference Diagram Q6A (Selected soil profiles)

Marks

Question 6 (continued)

(ii) Study Reference Diagram Q6B.

Explain how the major soil forming factors shown in the diagram have contributed to the formation of your chosen soil profile. **4**

Reference Diagram Q6B (Main factors affecting soil formation)

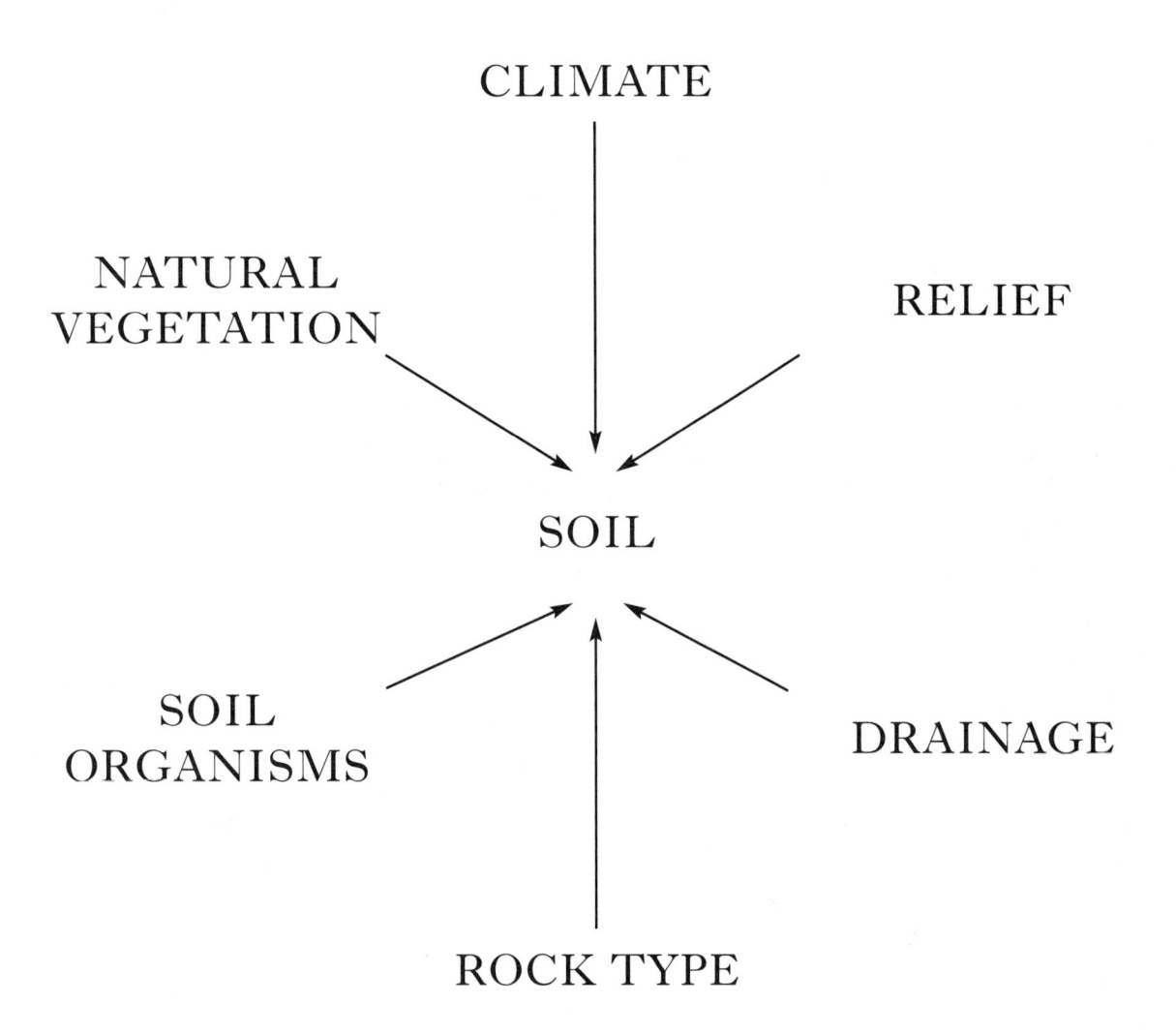

CLIMATE

NATURAL VEGETATION   RELIEF

SOIL

SOIL ORGANISMS   DRAINAGE

ROCK TYPE

[Turn over

SECTION C: Answer ONE question from this section,
ie either Question 7 or Question 8.

Marks

Question 7: Rural Geography

(a) **Describe** the main characteristics of **shifting cultivation**. 3

(b) *"In Central America, population density and loss of rainforest cover are closely related. During the last two decades, human activities have caused the deforestation of more than 120 000 square kilometres each year."*

Referring to a named area where shifting cultivation is carried out, **explain** the **impact** which deforestation and increased population density have had on the environment and way of life of the shifting cultivators. 4

**DO NOT ANSWER THIS QUESTION IF YOU HAVE
ALREADY ANSWERED QUESTION 7**

Question 8: Urban Geography

Study OS Map Extract number 1560/150: Worcester (*separate item*), and Reference Map Q8.

(a) **Describe** the urban environment of Area A and **explain** its location. 4

(b) For **either** Area B **or** Area C, **explain** the advantages of the residential environment. 3

Question 8 (continued)

Reference Map Q8 (Location of urban areas in Worcester)

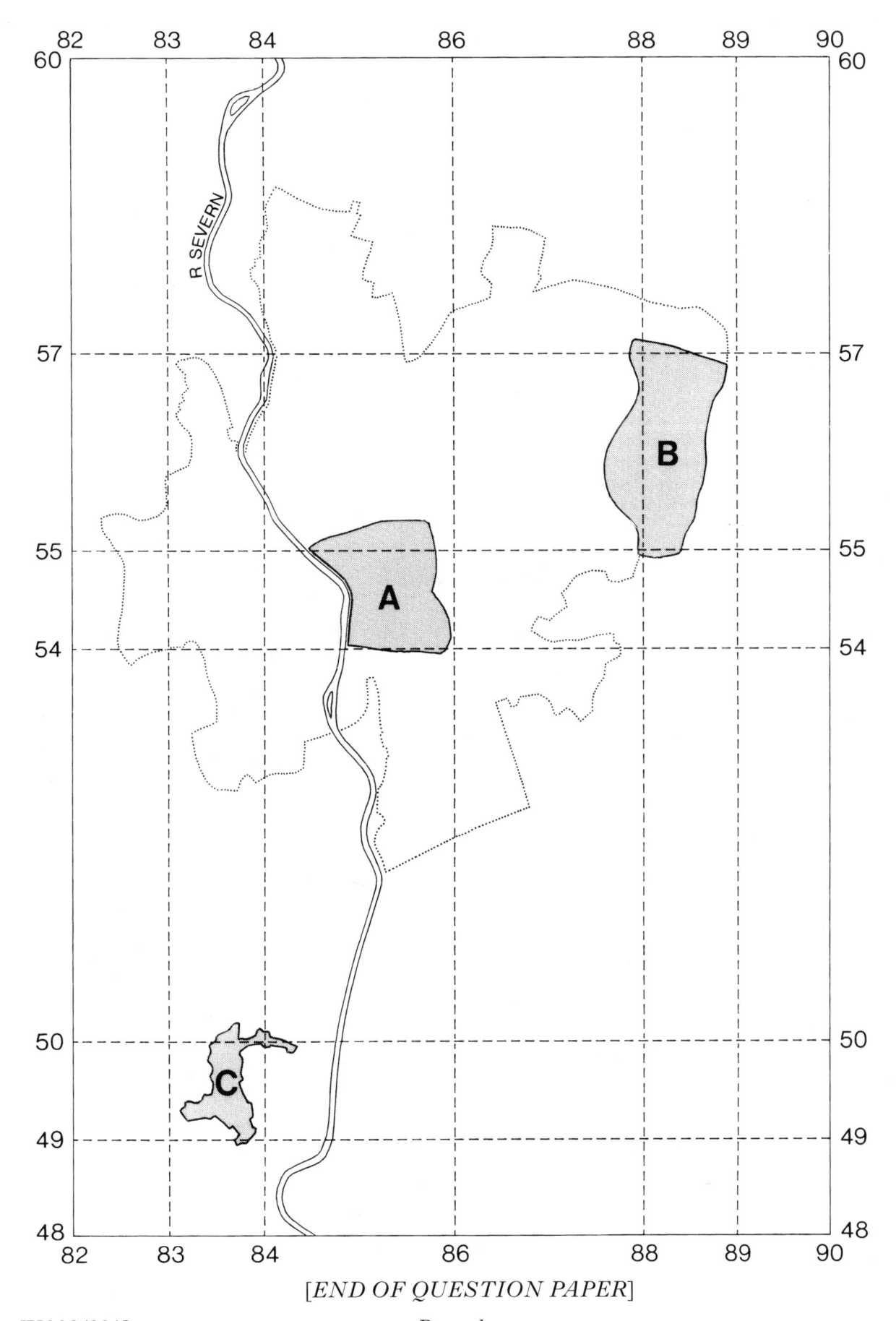

[END OF QUESTION PAPER]

[BLANK PAGE]

X208/303

NATIONAL
QUALIFICATIONS
2007

MONDAY, 28 MAY
10.50 AM – 12.05 PM

GEOGRAPHY
HIGHER
Paper 2
Environmental
Interactions

Two questions should be attempted, namely:

one question from **Section 1** (Questions 1, 2, 3) and
one question from **Section 2** (Questions 4, 5, 6).

Write the numbers of the **two** questions you have attempted in the marks grid on the back cover of your answer booklet.

The value attached to each question is shown in the margin.

Credit will be given for appropriate maps and diagrams, and for reference to named examples.

Questions should be answered in sentences.

Note The reference maps and diagrams in this paper have been printed in black only: no other colours have been used.

Marks

SECTION 1

You must answer ONE question from this Section.

Question 1 (Rural Land Resources)

(a) Study Reference Maps Q1A and Q1B.

Describe and **suggest reasons for** the location of Britain's National Parks. **5**

(b) Study Reference Diagram Q1 on *Page four*.

"*Tourism can bring benefits but also causes problems for National Parks.*"

With the aid of Reference Diagram Q1 and referring to a specific National Park **or** other named upland **or** coastal landscape area which you have studied:

(i) **describe** some of the benefits which an influx of tourists has brought; and

(ii) **suggest** and **evaluate** ways in which the **negative** effects of tourism can be tackled. **10**

(c) With the aid of annotated diagrams, **describe** and **explain** the formation of the main features of any **coastal landscapes** which you have studied. You should refer to erosional **and** depositional features in your answer. **10**

(25)

Question 1 — continued

Reference Map Q1A
(National Parks in Great Britain)

Reference Map Q1B
(Relief map of Great Britain)

[Turn over

Question 1 — continued

Reference Diagram Q1 (Positive and negative effects of tourism)

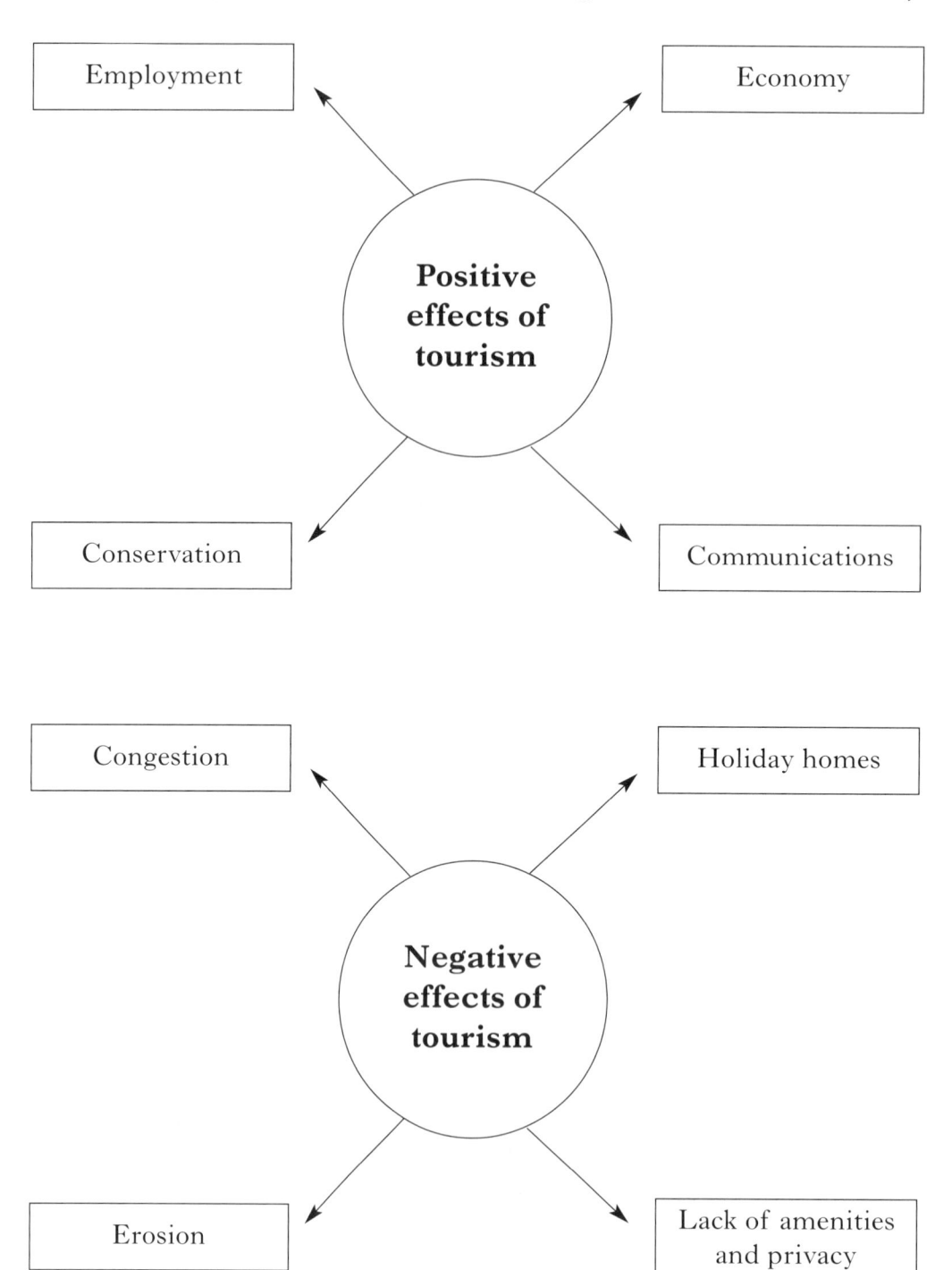

Positive effects of tourism:
- Employment
- Economy
- Conservation
- Communications

Negative effects of tourism:
- Congestion
- Holiday homes
- Erosion
- Lack of amenities and privacy

Marks

Question 2 (Rural Land Degradation)

(a) Study Reference Table Q2.

Describe the processes of soil erosion by both water and wind. 5

(b) **Describe** and **explain** the main human causes of land degradation in North America **and either** Africa north of the Equator **or** the Amazon Basin. 8

(c) Referring to named locations in **either** Africa north of the Equator **or** the Amazon Basin, **describe** the social and economic impact of land degradation on the people. 5

(d) Study Reference Diagram Q2.

For any **four** methods of soil conservation, **explain** how each helps to conserve soil and reduce land degradation. 7

(25)

Reference Table Q2 (Types of water and wind erosion)

Soil Erosion by Water	Soil Erosion by Wind
Rainsplash	Suspension
Sheet wash	Saltation
Rill erosion	Surface creep
Gully erosion	

Reference Diagram Q2 (Soil conservation strategies)

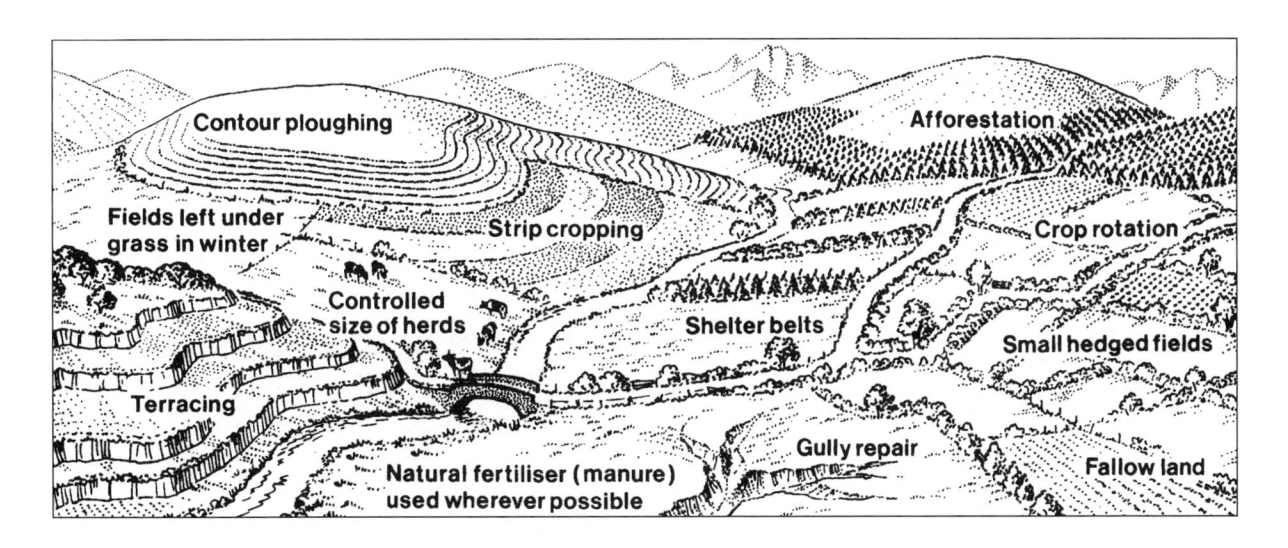

Contour ploughing; Afforestation; Fields left under grass in winter; Strip cropping; Crop rotation; Controlled size of herds; Shelter belts; Small hedged fields; Terracing; Gully repair; Fallow land; Natural fertiliser (manure) used wherever possible

[Turn over

Marks

Question 3 (River Basin Management)

(a) Study Reference Maps Q3A, Q3B and Q3C.

For Asia **or** Africa **or** North America, **describe** and **explain** the general distribution of the main river basins. 5

(b) "*Among the 30 largest dams planned for the Narmada river in India, the Sardar Sarovar is the largest. Most of the water held back by the dam will be used in the neighbouring state of Gujarat as it expands its rice and cotton production.*"

Study Reference Maps Q3D, Q3E and Reference Diagram Q3 on *Page eight*.

Explain why there is a need for water management in the Narmada River Basin and in Gujarat State. 5

(c) For the Narmada River Project **or** any other river basin management project in Asia **or** Africa **or** North America, **explain** the political problems that may have resulted from the project. 3

(d) **Describe** and **suggest reasons for** the social, economic and environmental benefits **and** adverse consequences of a named water control project in Asia **or** Africa **or** North America. 12

(25)

Question 3 — continued

Reference Map Q3A
(Major river basins of Africa)

Reference Map Q3B
(Major river basins of North America)

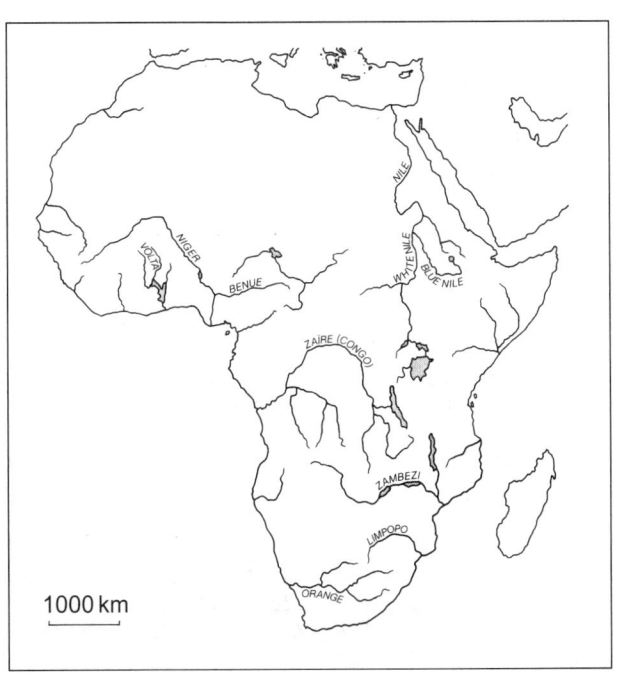

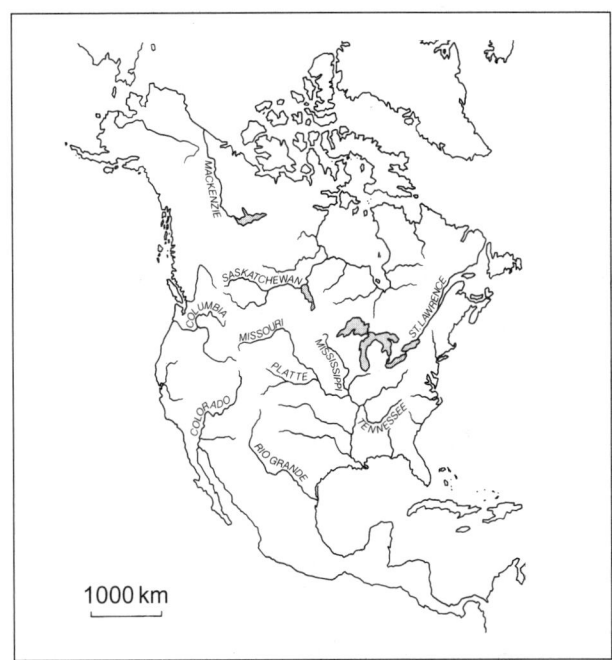

Reference Map Q3C (Major river basins of Asia)

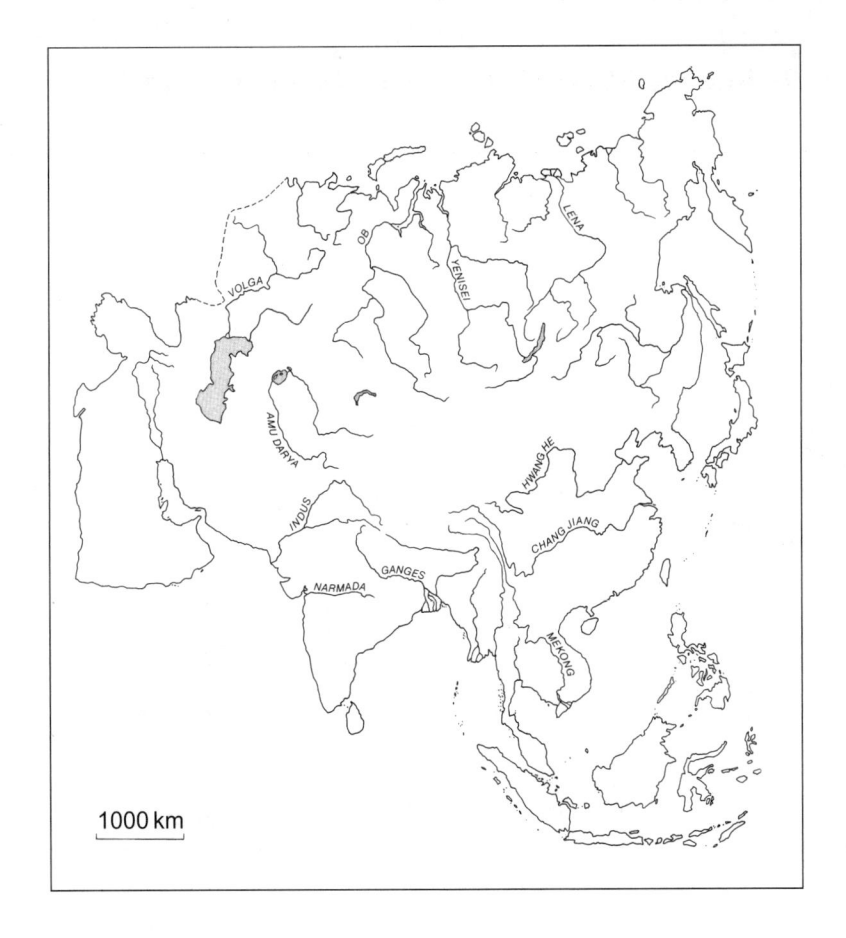

Question 3 — continued

Reference Map Q3D
(Location of Narmada River Basin in India)

Reference Diagram Q3
(Climate graphs for Ahmedabad)

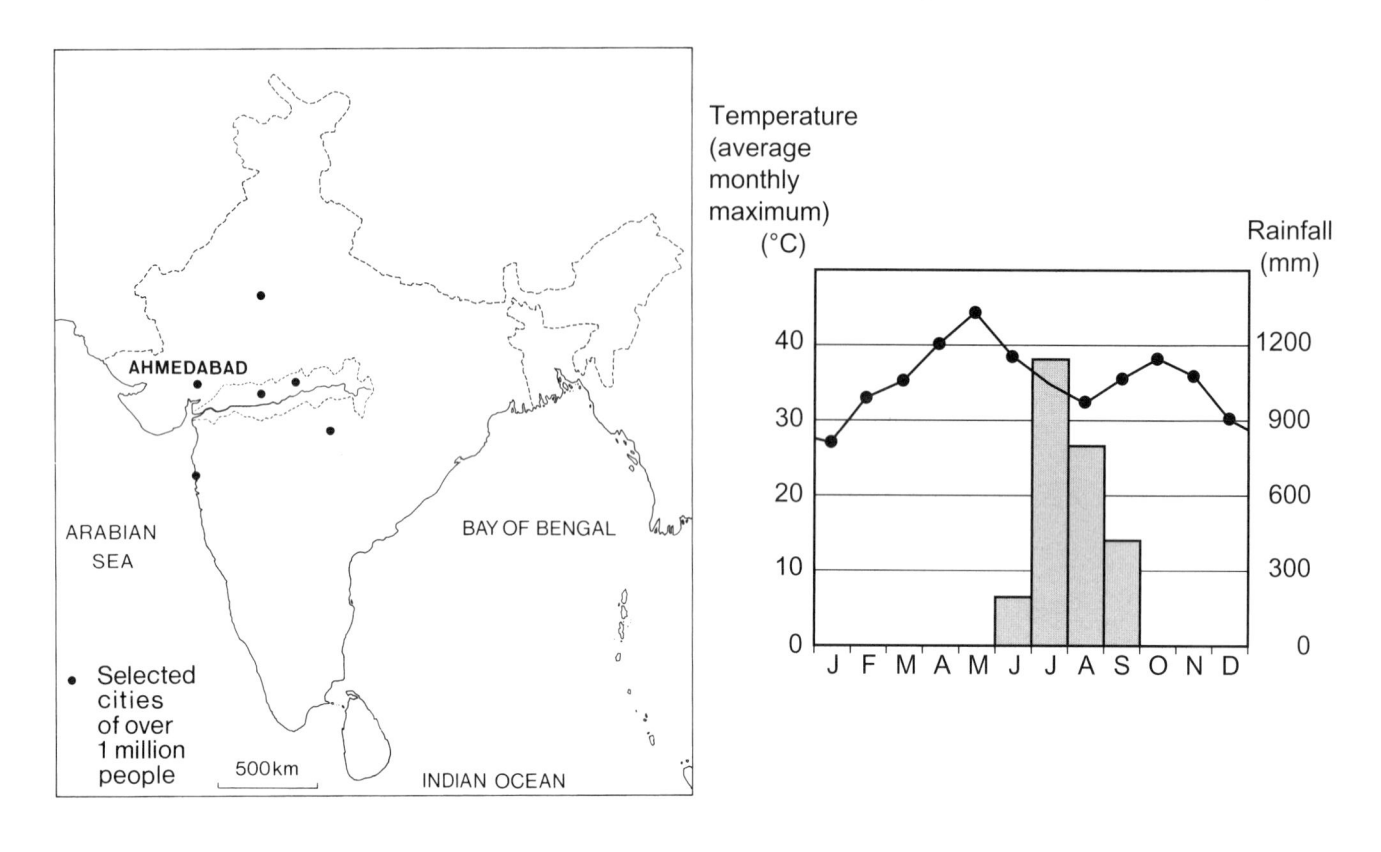

Reference Map Q3E (Basin of Narmada River)

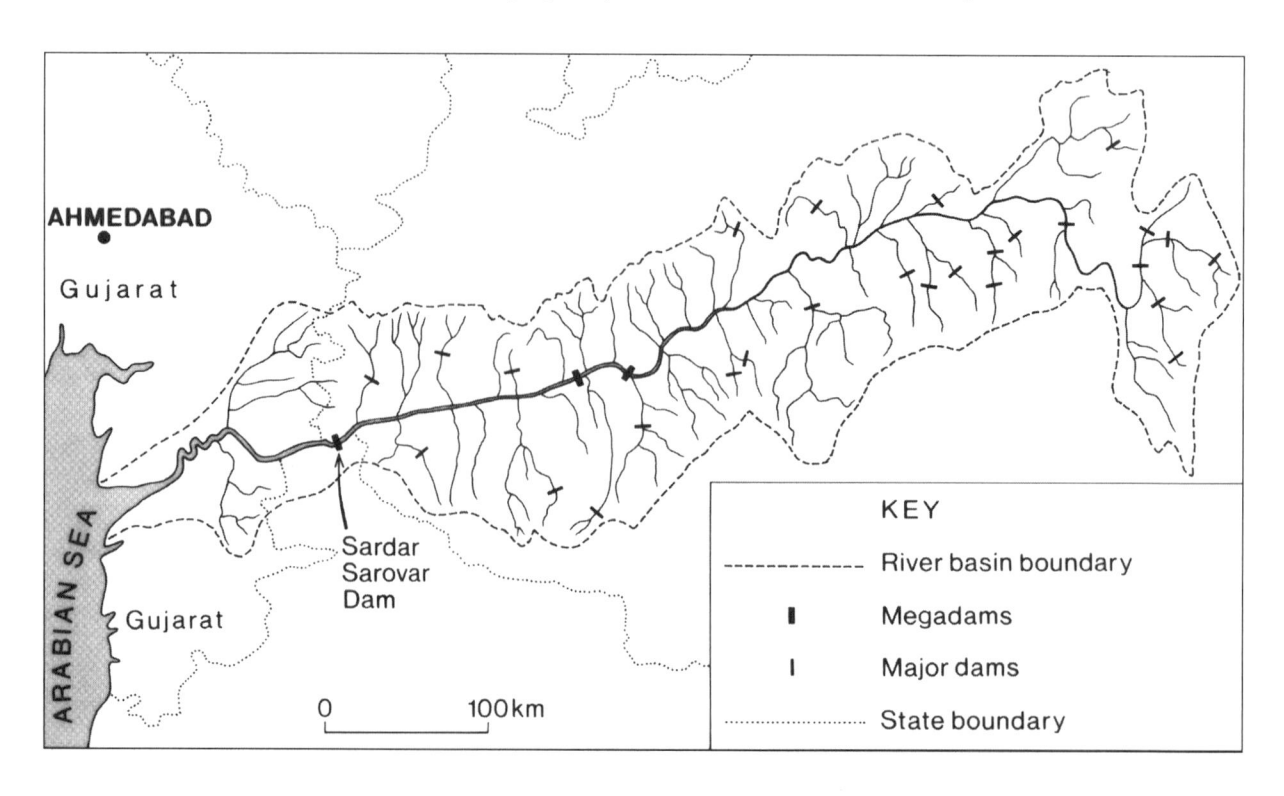

[Turn over for Question 4 on *Page ten*

Marks

SECTION 2

You must answer ONE question from this Section.

Question 4 (Urban Change and its Management)

Study Reference Maps Q4A and Q4B.

(*a*) **Describe** and **suggest reasons** for the changing distribution of the world's largest urban areas over the last 50 years.   **6**

(*b*) With reference to a named city which you have studied in an **ELDC** (Economically Less Developed Country):

  (i) **describe** the social, economic and environmental problems which have resulted from its rapid growth;

  (ii) **describe** some of the methods used to tackle these problems; and

  (iii) **comment** on the effectiveness of the methods used.   **10**

(*c*) The rural-urban fringe of many cities in **EMDCs** (Economically More Developed Countries) is frequently an area of land-use conflict.

Study Reference Map Q4C on *Page twelve*.

Referring to Edinburgh **or** any other named city which you have studied in an **EMDC**:

  (i) **suggest** why land-use conflicts may have arisen; and

  (ii) **comment** on the effectiveness of strategies such as the creation of Green Belts, in resolving these conflicts.   **9**

(25)

Question 4 - continued

Reference Map Q4A (Ten largest urban areas in the world in 1957)

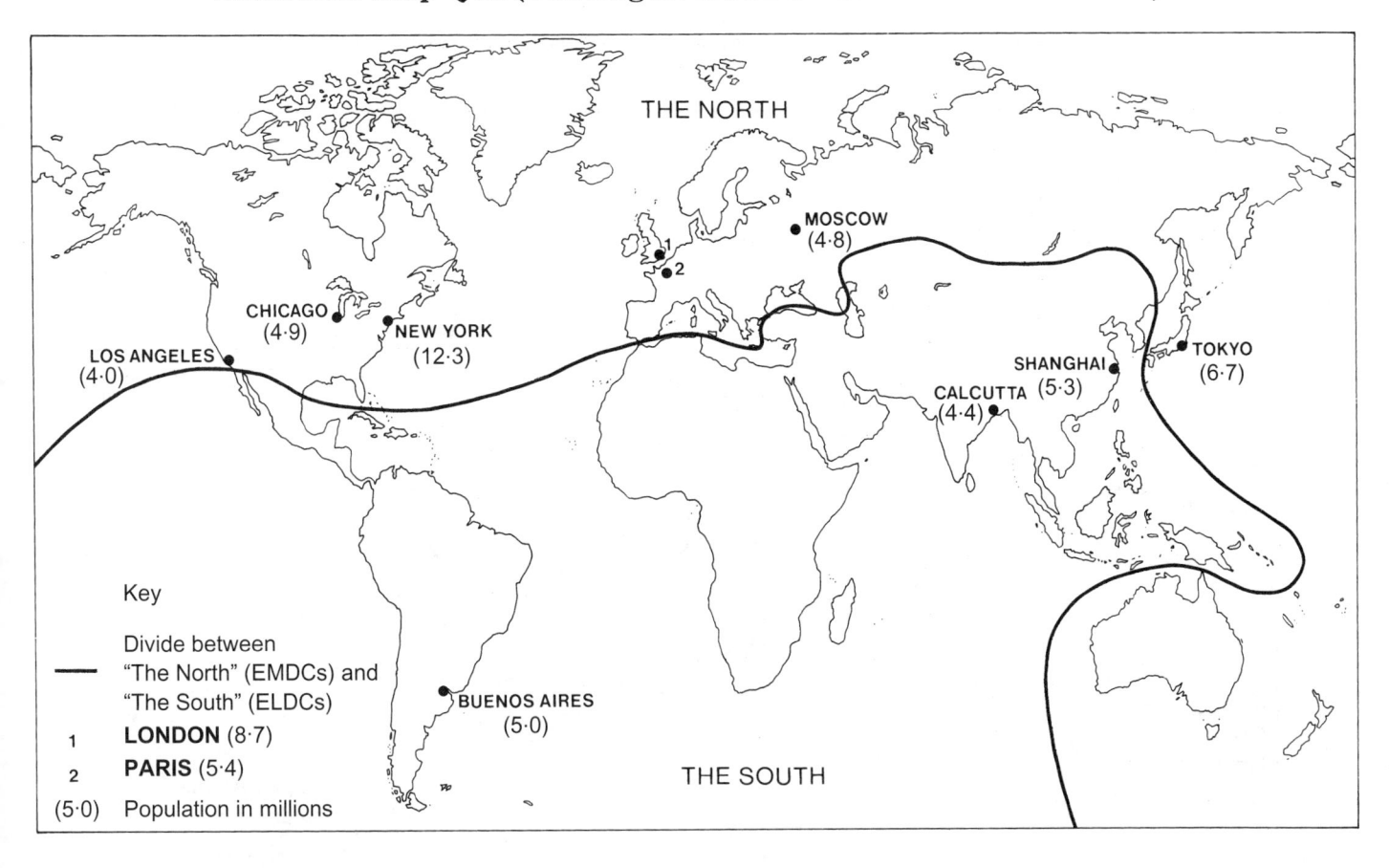

Key
— Divide between "The North" (EMDCs) and "The South" (ELDCs)
1 **LONDON** (8·7)
2 **PARIS** (5·4)
(5·0) Population in millions

THE NORTH
MOSCOW (4·8)
CHICAGO (4·9)
NEW YORK (12·3)
LOS ANGELES (4·0)
SHANGHAI (5·3)
CALCUTTA (4·4)
TOKYO (6·7)
BUENOS AIRES (5·0)
THE SOUTH

Reference Map Q4B (Ten largest urban areas in the world in 2007)

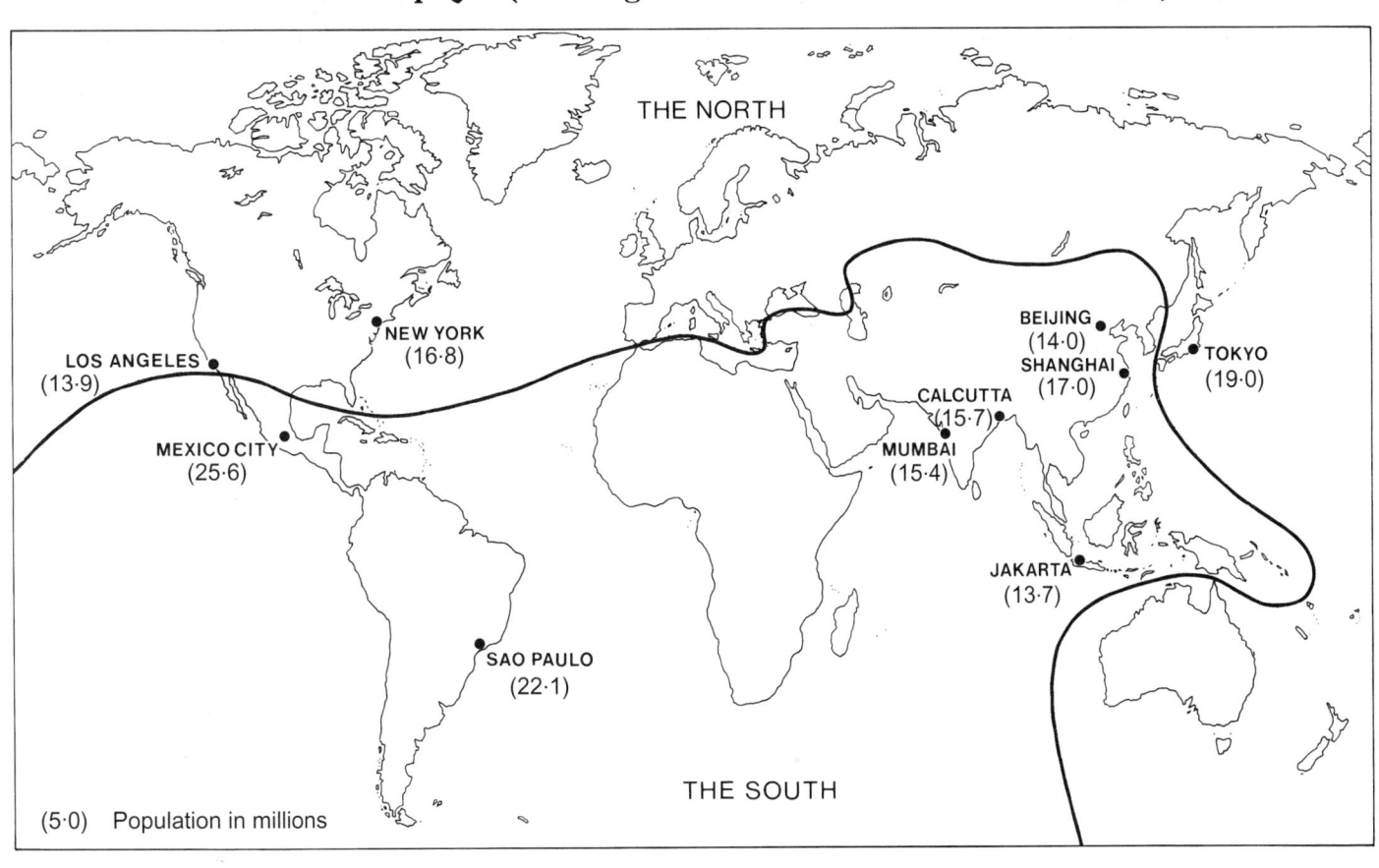

THE NORTH
NEW YORK (16·8)
LOS ANGELES (13·9)
MEXICO CITY (25·6)
BEIJING (14·0)
SHANGHAI (17·0)
CALCUTTA (15·7)
MUMBAI (15·4)
TOKYO (19·0)
JAKARTA (13·7)
SAO PAULO (22·1)
THE SOUTH

(5·0) Population in millions

Question 4 - continued

Reference Map Q4C (Pressures on Edinburgh's Green Belt)

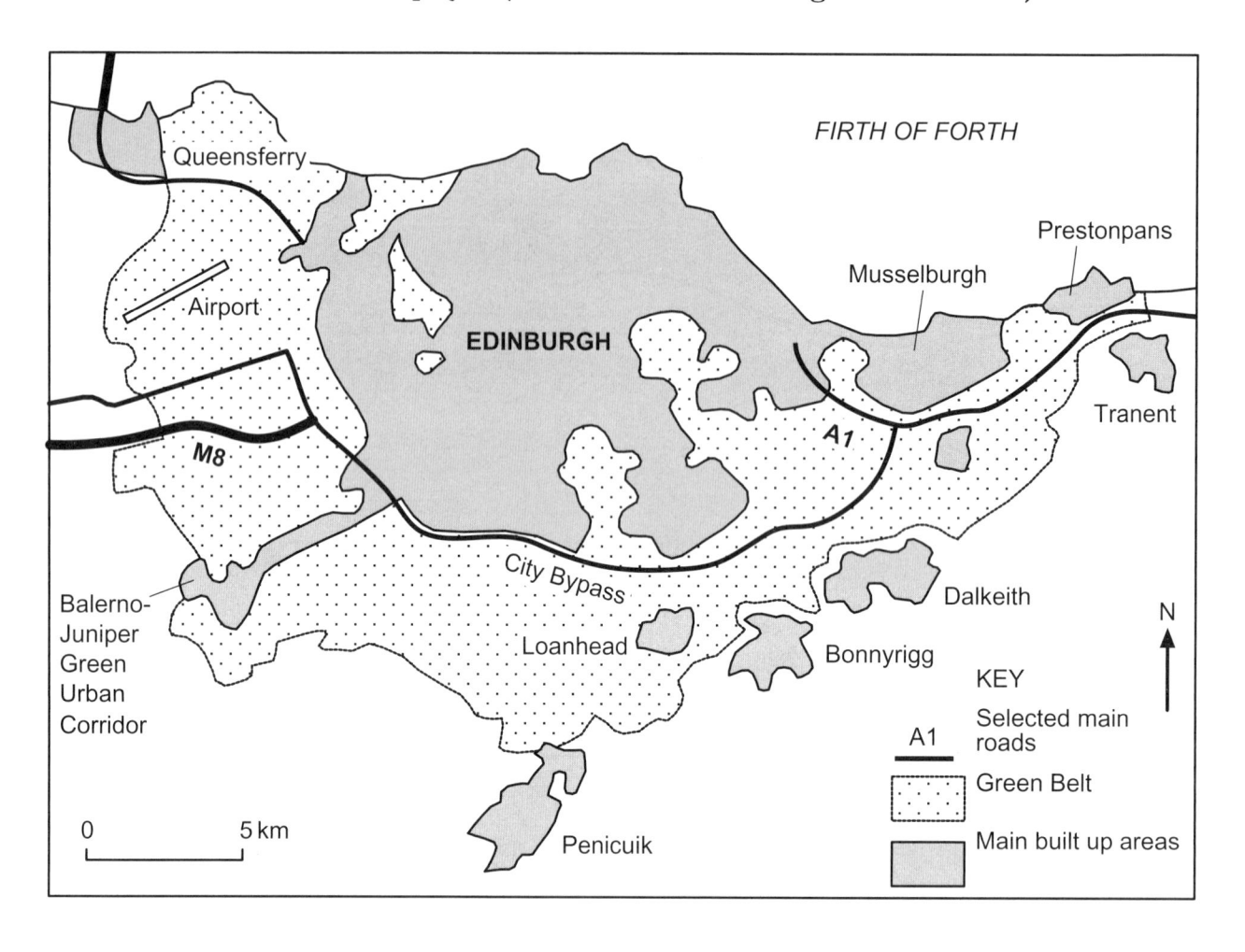

[Turn over for Question 5 on *Page fourteen*

Question 5 (European Regional Inequalities)

(a) Study Reference Table Q5A and Reference Map Q5A.

Describe and **suggest reasons** for the differences in levels of development between the countries shown.

Marks

5

(b) Study Reference Map Q5B on *Page sixteen* and Reference Table Q5B.

To what extent does the data provide evidence of regional inequalities within Poland?

6

(c) Many countries within the European Union (EU) have marked regional inequalities.

For a named country in the EU, select **one** less developed region and:

(i) **describe** and **account for** the social and economic problems faced by the region;

6

(ii) **outline** the efforts being made by both the national government and European Union agencies to tackle the problems, and comment on their effectiveness.

8

(25)

Reference Table Q5A (Selected EU country statistics)

	France	Spain	Poland
Population (millions)	60	41	39
GDP ($US per capita)	26 920	21 460	10 560
Life expectancy	79	79	74
Health expenditure ($US per capita)	2567	1607	629
Urban %	76	76	62
Date of EU membership	1957	1986	2004

Data: UN World Development Handbook

Question 5 — continued

Reference Map Q5A (Selected European Union countries)

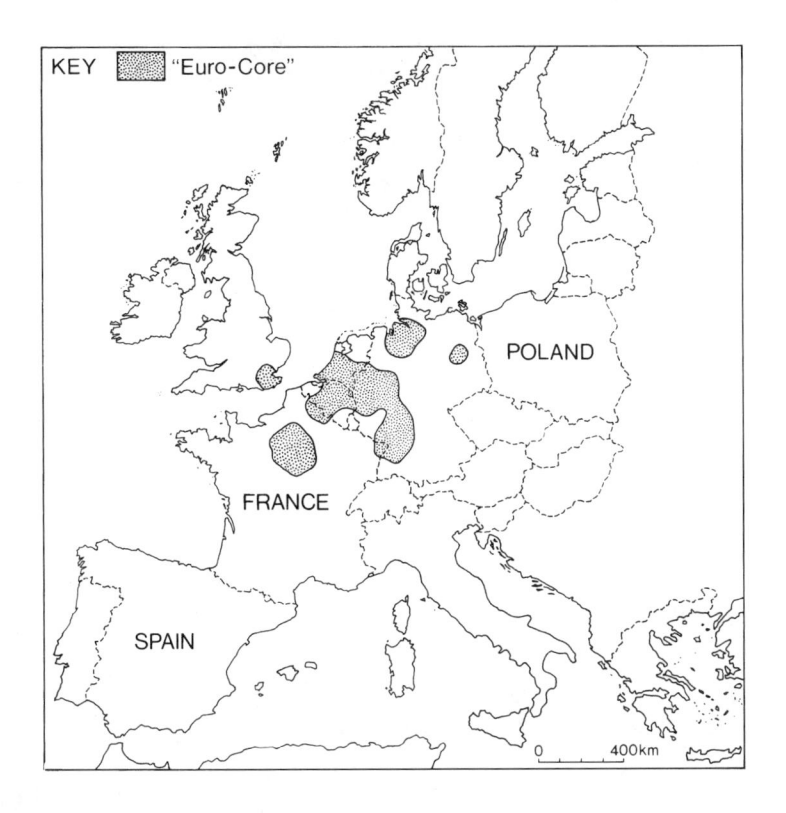

Reference Table Q5B (Polish Provinces—selected statistics)

Province	GDP per capita (in 1000 zloty) 2002	Private vehicles % national total 2003	Electricity production % total 2003
Lower Silesia	21·2	7	8
Cuiavia & Pomerania	18·6	5	2
Lublin	14·3	5	1
Lubusz	17·8	3	<1
Lódz´	18·5	7	20
Lesser Poland	17·7	8	6
Mazovia	31·1	15	13
Opole	16·7	3	6
Sub-Carpathia	14·6	5	2
Podlassia	15·7	3	<1
Pomerania	20·3	6	2
Silesia	22·6	12	20
Kielce	16·0	3	5
Varmia & Masuria	15·2	3	<1
Greater Poland	21·0	11	10
Western Pomerania	20·2	4	4

[Turn over to see map of Provinces

Question 5 — continued

Reference Map Q5B (Poland: Provinces)

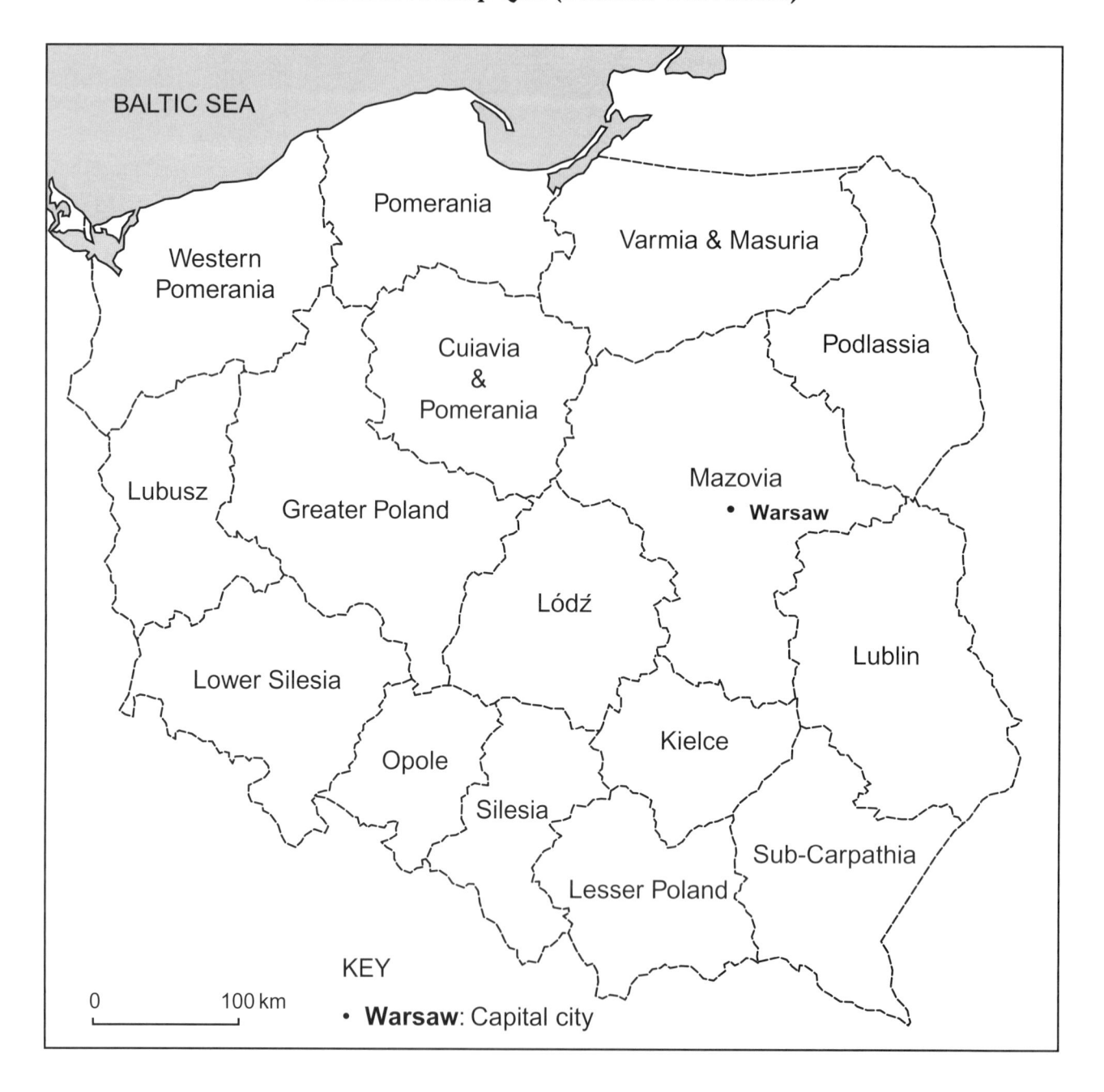

[Turn over for Question 6 on *Page eighteen*

Question 6 (Development and Health)

(a) % Adult literacy is a social indicator of development. Identify one other social indicator of development **and** one economic indicator of development. For each indicator you have identified, **explain** how it might illustrate a country's level of development. 4

(b) Study Reference Table Q6.

Reference Table Q6 (Adult literacy rates in selected Economically Less Developed Countries (ELDCs))

Country	% Adult Literacy
Afghanistan	35
Bolivia	85
Burkina Faso	22
Cuba	96
Kenya	81
Malaysia	87
Sri Lanka	91

The table shows that there are considerable differences in levels of development between Economically Less Developed Countries (ELDCs). Referring to these countries and/or to other ELDCs you have studied, **suggest reasons** why such differences exist **between** countries. 5

(c) Many ELDCs have marked differences in levels of development **within** their borders. For a named ELDC, **explain** the differences found **within** the country. 4

(d) Study Reference Map Q6 which shows the main areas of the world at risk from cholera.

Referring to cholera **or** malaria **or** bilharzia/schistosomiasis:

(i) **describe** the physical and human factors which put people at risk of contracting the disease;

(ii) **describe** and **explain** the strategies used in controlling the spread of the disease; and

(iii) **explain** the benefits to ELDCs of controlling the disease. 12

(25)

Question 6 — continued

Reference Map Q6 (Countries with a recent cholera outbreak)

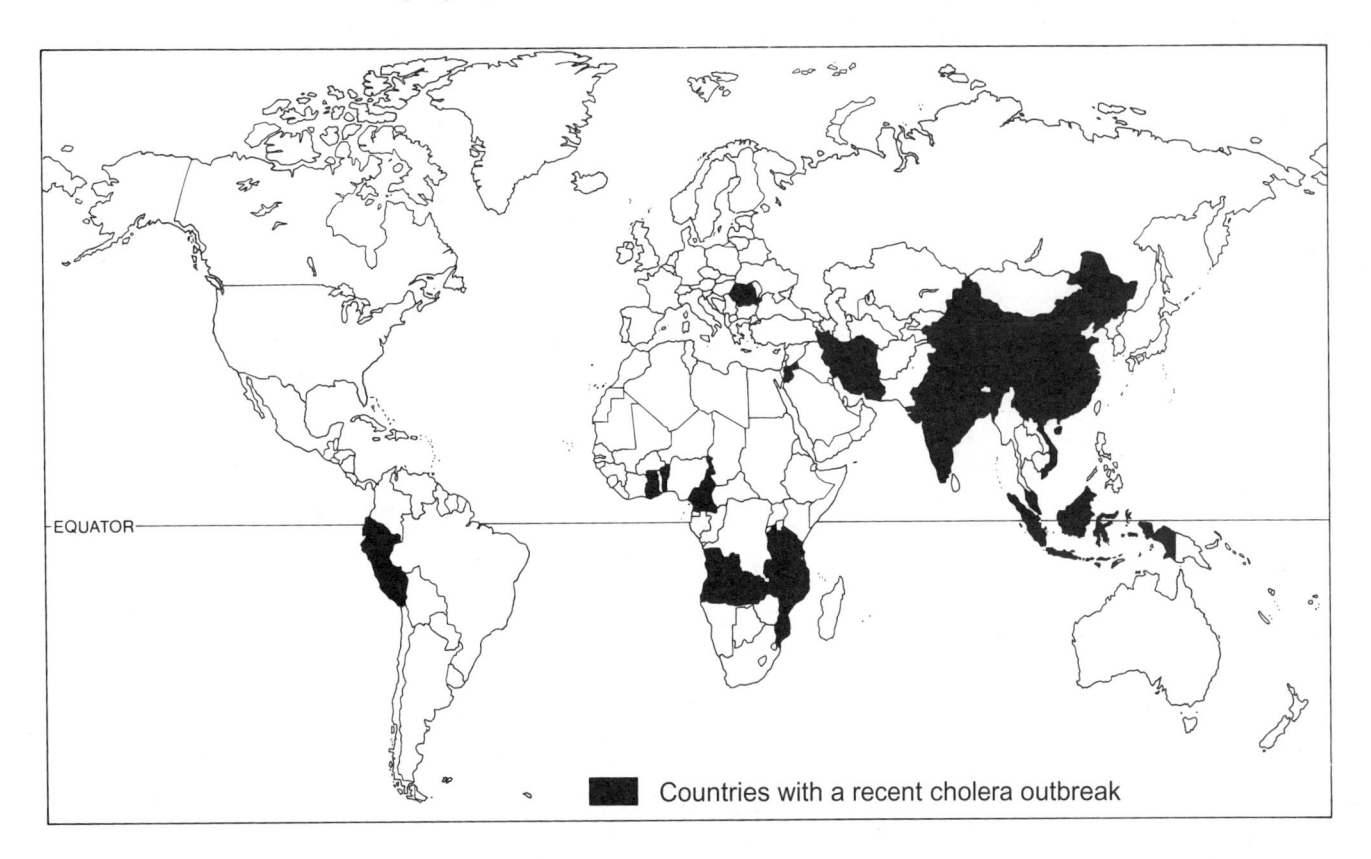

[BLANK PAGE]

HIGHER
2008

[BLANK PAGE]

X208/301

NATIONAL
QUALIFICATIONS
2008

THURSDAY, 22 MAY
9.00 AM – 10.30 AM

GEOGRAPHY
HIGHER
Paper 1
Physical and
Human Environments

Six questions should be attempted, namely:

all four questions in **Section A** (Questions 1, 2, 3 and 4);

one question from **Section B** (Question 5 **or** Question 6);

one question from **Section C** (Question 7 **or** Question 8).

Write the numbers of the **six** questions you have attempted in the marks grid on the back cover of your answer booklet.

The value attached to each question is shown in the margin.

Credit will be given for appropriate maps and diagrams, and for reference to named examples.

Questions should be answered in sentences.

Note The reference maps and diagrams in this paper have been printed in black only: no other colours have been used.

Extract No 1659/EXP-OL15

1:25 000 Scale
Explorer Series

Four colours should appear above; if not then please return to the invigilator.
Four colours should appear above; if not then please return to the invigilator.

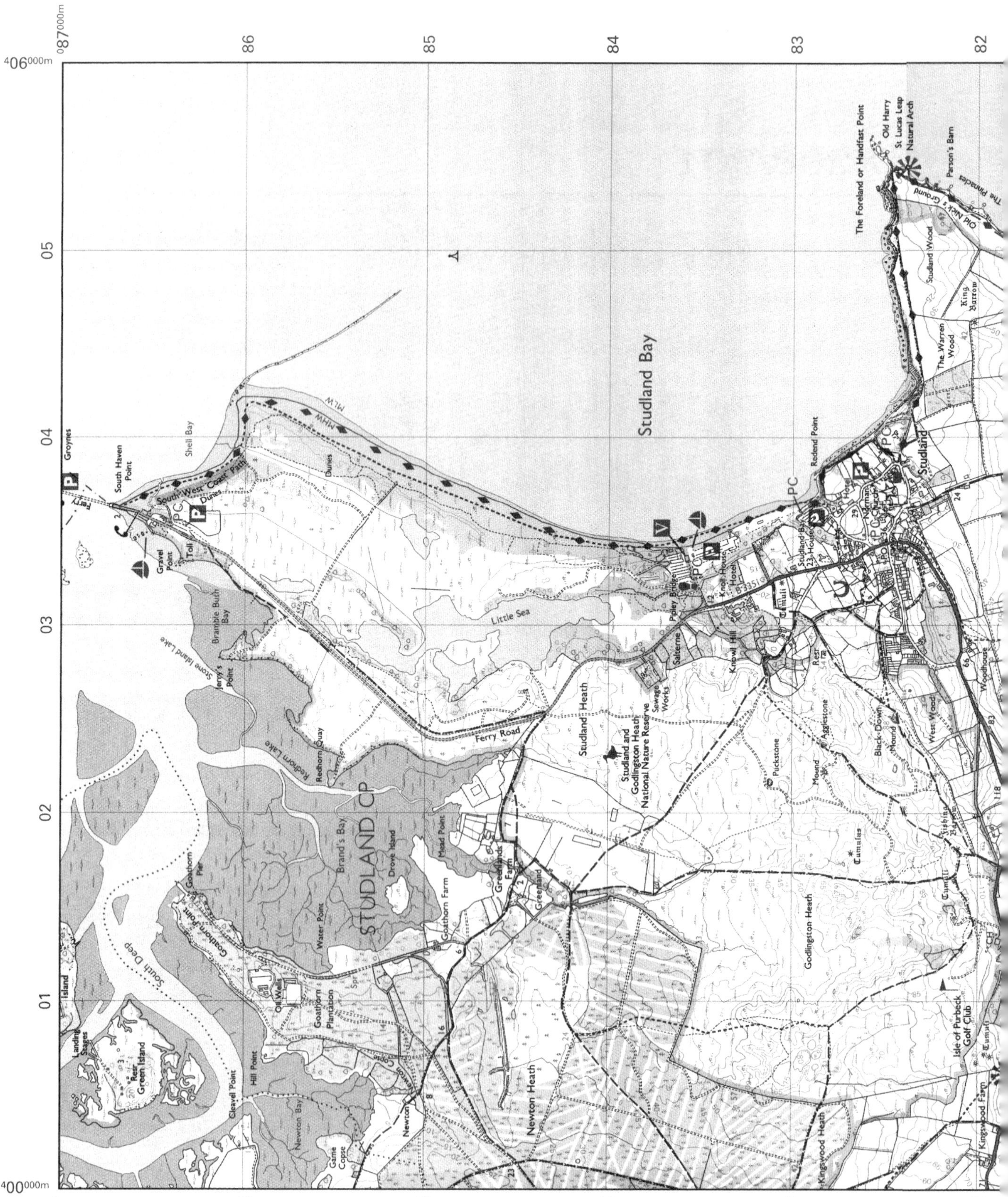

Extract produced by Ordnance Survey 2007.

042, 786

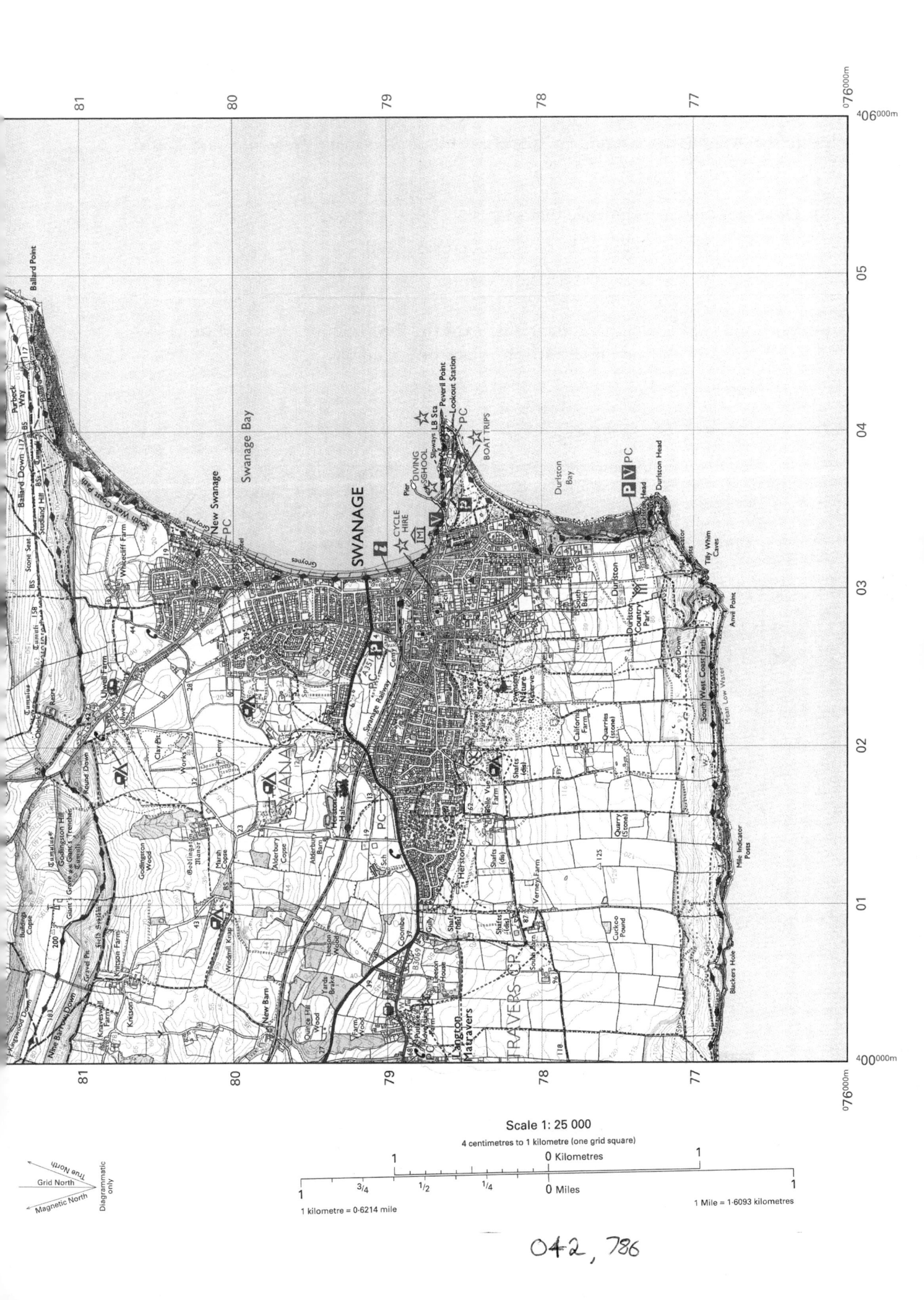

Marks

SECTION A: Answer ALL questions in this section

Question 1: Lithosphere

Study OS Map Extract number 1659/Exp–OL15: Swanage (*separate item*), **and** Reference Map Q1.

(*a*) **Describe** the map evidence that shows:

   (i) Areas A and B are areas of coastal erosion, and

   (ii) Area C is an area of coastal deposition.

12

(*b*) With the aid of annotated diagrams, **explain** the various stages and processes involved in the formation of **either** a stack **or** a sand bar.

8

Reference Map Q1

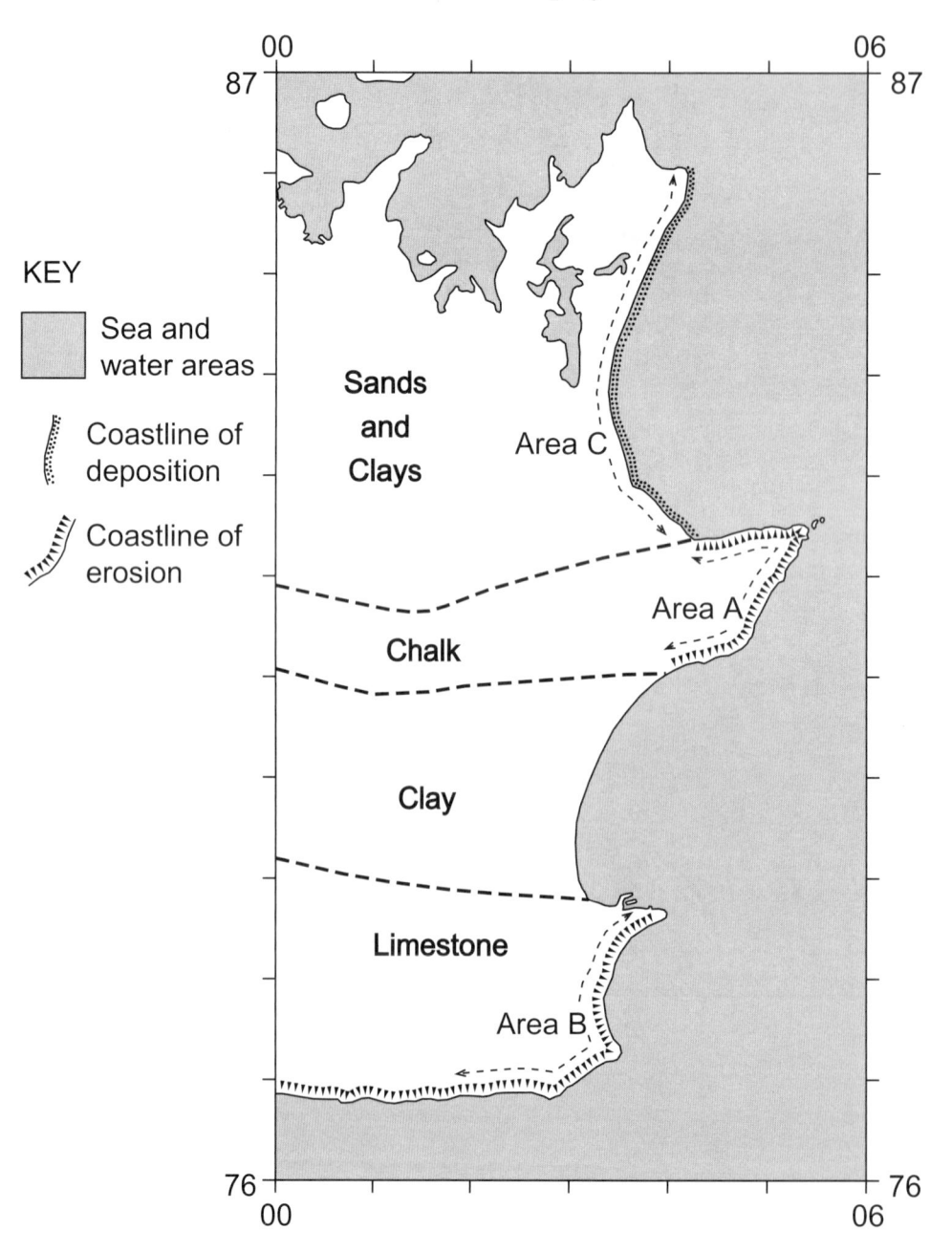

Question 2: Biosphere

Study Reference Diagram Q2.

Describe and **give reasons for** the changes in plant types likely to be observed across the transect as you move inland from the coast.

You should refer to named plant species likely to be found growing at different sites and to influencing factors **such as** shelter, pH and distance from the sea.

16

Reference Diagram Q2 (Transect across sand dune coastline)

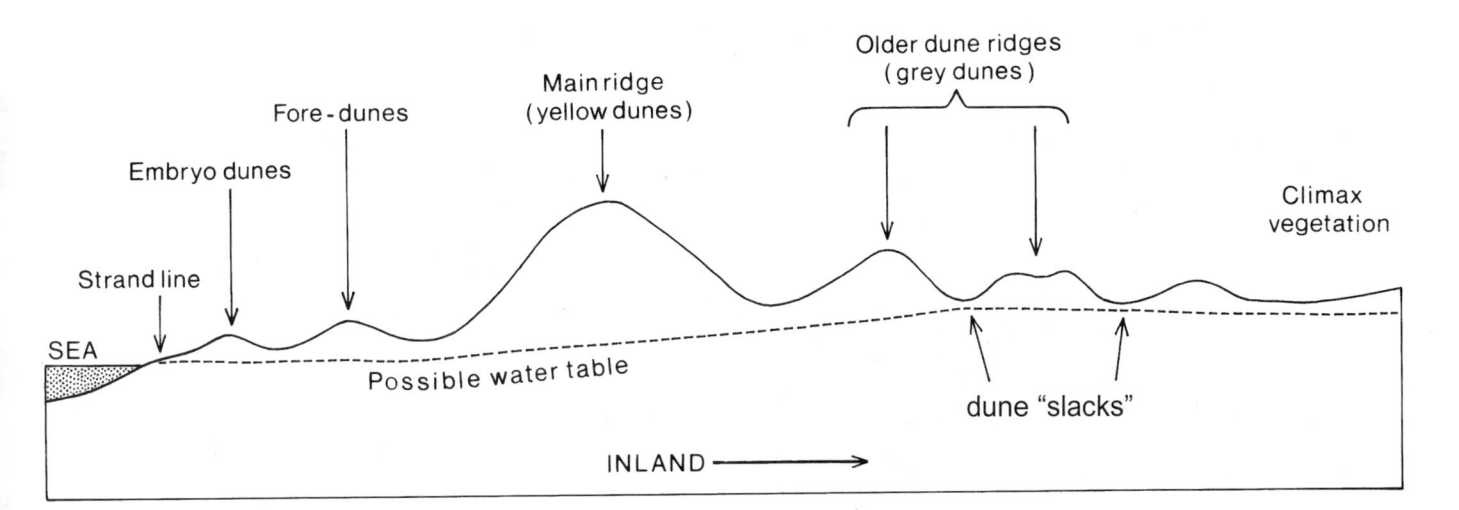

[Turn over

Question 3: Population Geography

Study Reference Diagram Q3 which shows the five stages of the Model of Demographic Transition.

(a) **Describe** and **explain** the changes in the **total population** in stages 1, 2 and 3 of the model. **10**

(b) The total population levels off in stage 4 and starts to fall in stage 5.

Describe the problems which a government may face when a country is in stage 5. **8**

Reference Diagram Q3 (Model of demographic transition)

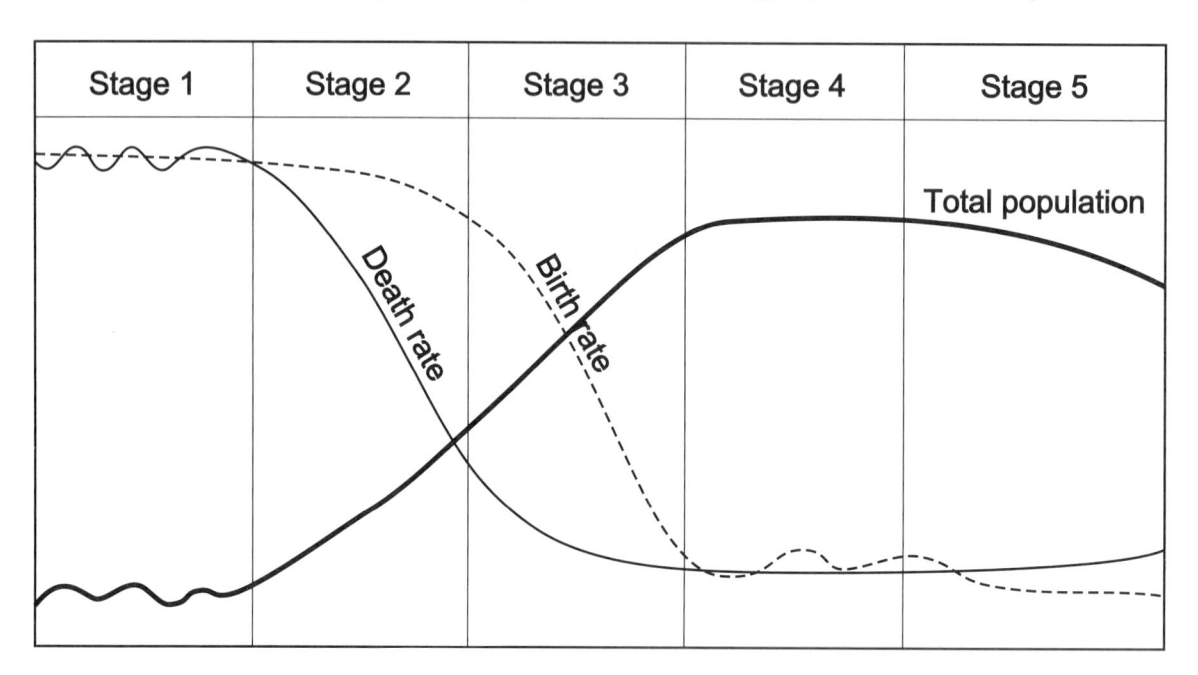

Question 4: Urban Geography

(a) For a named city which you have studied in an EMDC (Economically More Developed Country), **explain** the ways in which the site and situation have contributed to its growth. **8**

(b) Study Reference Photograph Q4.

"*Traffic congestion is now a major problem facing many cities in EMDCs*".

Describe and **explain** schemes which have been introduced to reduce problems of traffic management in any named city you have studied in an EMDC. **10**

Reference Photograph Q4 (Traffic congestion)

[Turn over

SECTION B: Answer ONE question from this section,
ie either Question 5 or Question 6.

Question 5: Atmosphere

Study Reference Diagram Q5.

Explain the physical **and** human factors that might have led to the changes in global air temperatures shown in the diagram. **14**

Reference Diagram Q5 (Global air temperatures 1855–2005)

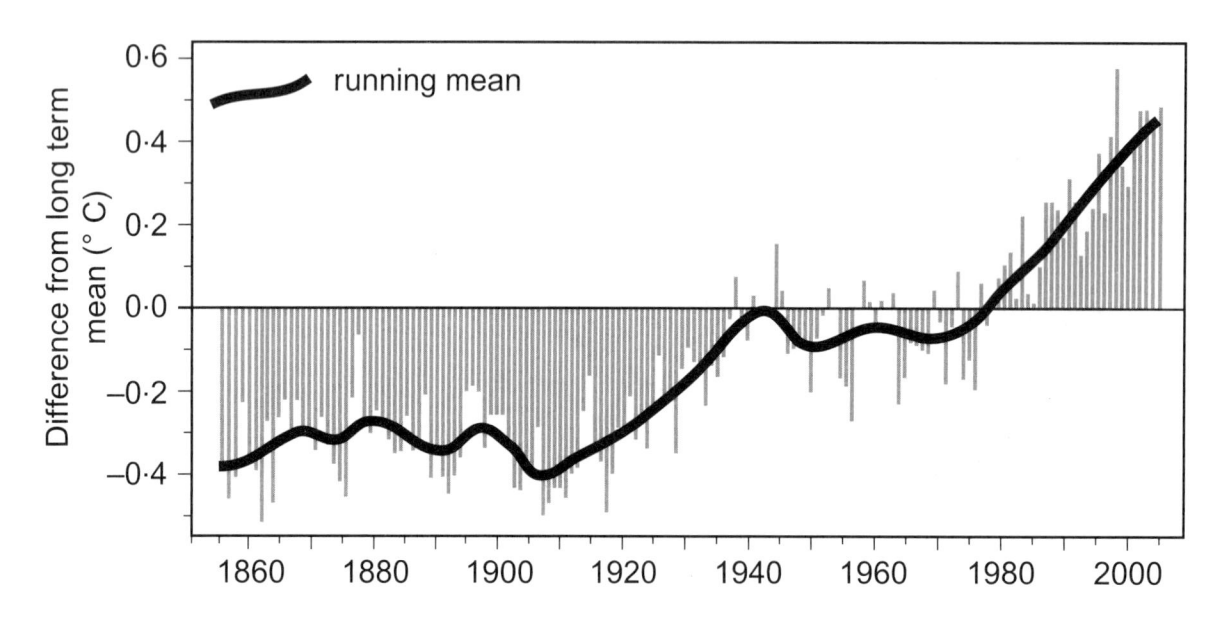

DO NOT ANSWER THIS QUESTION IF YOU HAVE ALREADY ANSWERED QUESTION 5

Question 6: Hydrosphere

(a) With the aid of a diagram, **describe** the global hydrological cycle. 6

(b) Study Reference Diagram Q6.

Explain the **differences** in discharge between the urban and rural hydrographs shown in the diagram following a heavy rain storm. 8

Reference Diagram Q6 (Flood hydrographs)

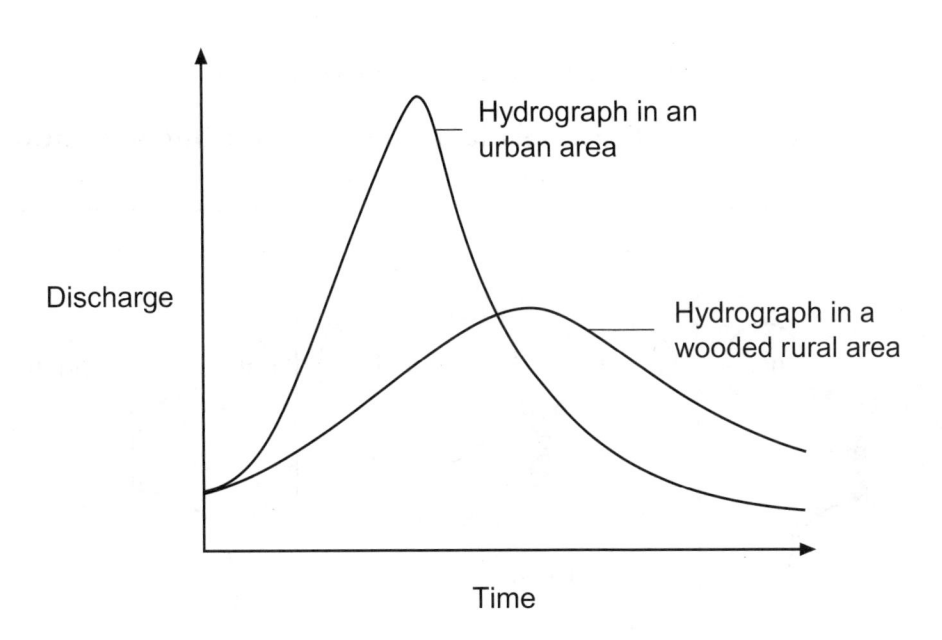

[Turn over

Marks

SECTION C: Answer ONE question from this section, ie either Question 7 or Question 8.

Question 7: Rural Geography

Study Reference Diagram Q7 which shows three different farming systems.

Choose **one** of these farming systems and:

(i) **explain** the ways in which the diagram reflects the main features of your chosen system; 6

(ii) referring to a **named** area where **your chosen system** is carried out, **describe** the changes in farming practices that have taken place in recent years. 8

Reference Diagram Q7 (Farming systems)

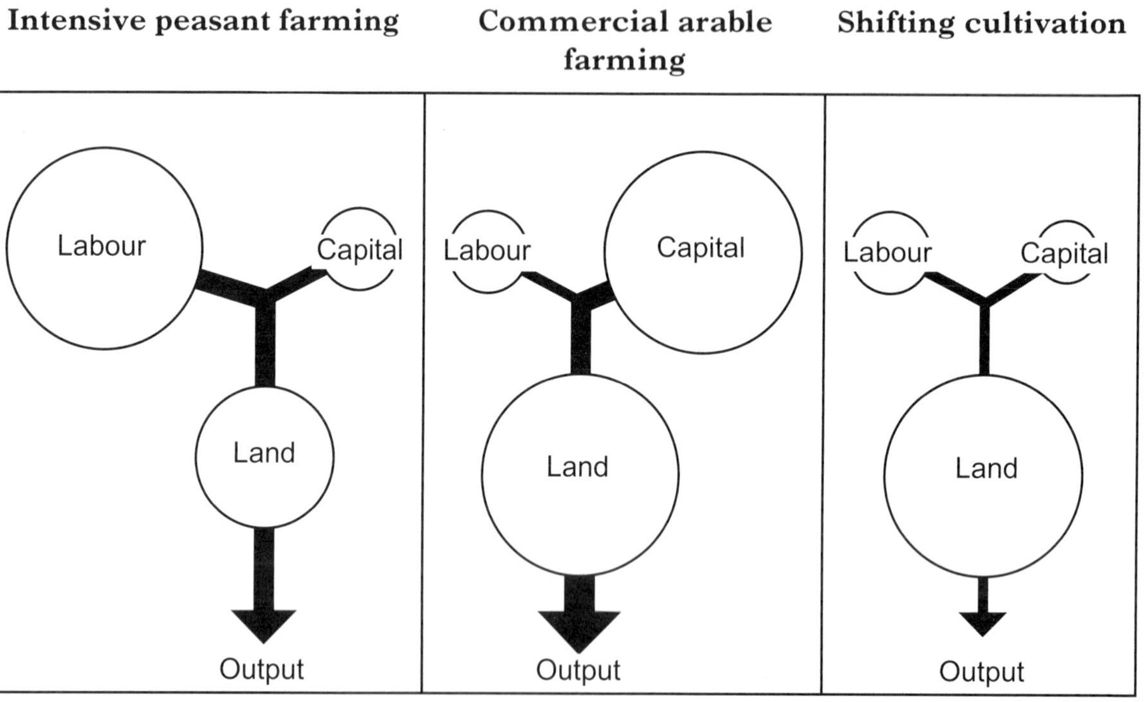
Intensive peasant farming
Commercial arable farming
Shifting cultivation
Labour
Capital
Land
Output
Labour
Capital
Land
Output
Labour
Capital
Land
Output

DO NOT ANSWER THIS QUESTION IF YOU HAVE ALREADY ANSWERED QUESTION 7

Marks

Question 8: Industrial Geography

(a) Study Reference Diagram Q8A.

   Describe and **explain** the impact of industry on the environment of an old industrial area such as that shown in Reference Diagram Q8A.

6

Reference Diagram Q8A (Old industrial landscape—South Wales)

(b) Study Reference Diagram Q8B.

   For South Wales, or any other industrial concentration in the EU, **describe** and **explain** the main location factors that influence the location of new industrial developments.

8

Reference Diagram Q8B (New industrial landscape—South Wales)

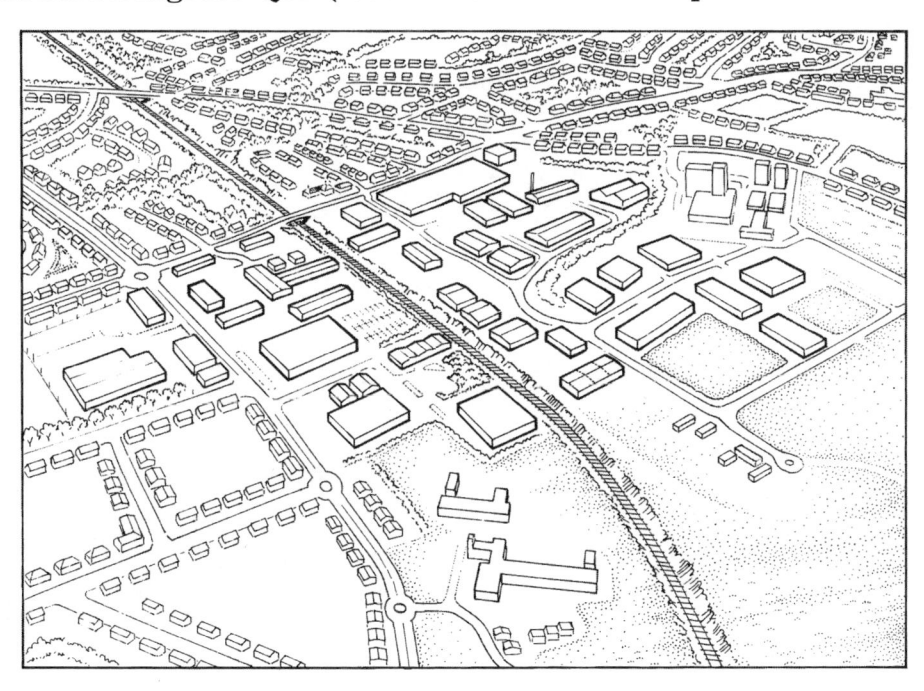

[END OF QUESTION PAPER]

X208/303

NATIONAL
QUALIFICATIONS
2008

THURSDAY, 22 MAY
10.50 AM – 12.05 PM

GEOGRAPHY
HIGHER
Paper 2
Environmental
Interactions

Two questions should be attempted, namely:

one question from **Section 1** (Questions 1, 2, 3) and
one question from **Section 2** (Questions 4, 5, 6).

Write the numbers of the **two** questions you have attempted in the marks grid on the back cover of your answer booklet.

The value attached to each question is shown in the margin.

Credit will be given for appropriate maps and diagrams, and for reference to named examples.

Questions should be answered in sentences.

Note   The reference maps and diagrams in this paper have been printed in black only: no other colours have been used.

SECTION 1

You must answer ONE question from this Section.

Question 1 (Rural Land Resources)

(a) The Peak District and Yorkshire Dales National Parks are two areas of Upland Limestone.

Study Reference Diagram Q1A.

Describe and **explain** the physical features associated with upland limestone landscapes. Both surface and underground features should be included in your answer. **20**

(b) For the Peak District National Park, **or** a named upland area you have studied:

(i) **describe** the opportunities which this landscape provides for a variety of land uses; and **8**

(ii) **explain** the environmental problems and conflicts which may arise from the competing demands of these different land uses. **14**

(c) Study Reference Diagram Q1B.

Select **one** of the conservation strategies and **explain** the ways in which it helps to protect the landscape. **8**

(50)

Question 1 — continued

Reference Diagram Q1A (Carboniferous Limestone Landscape)

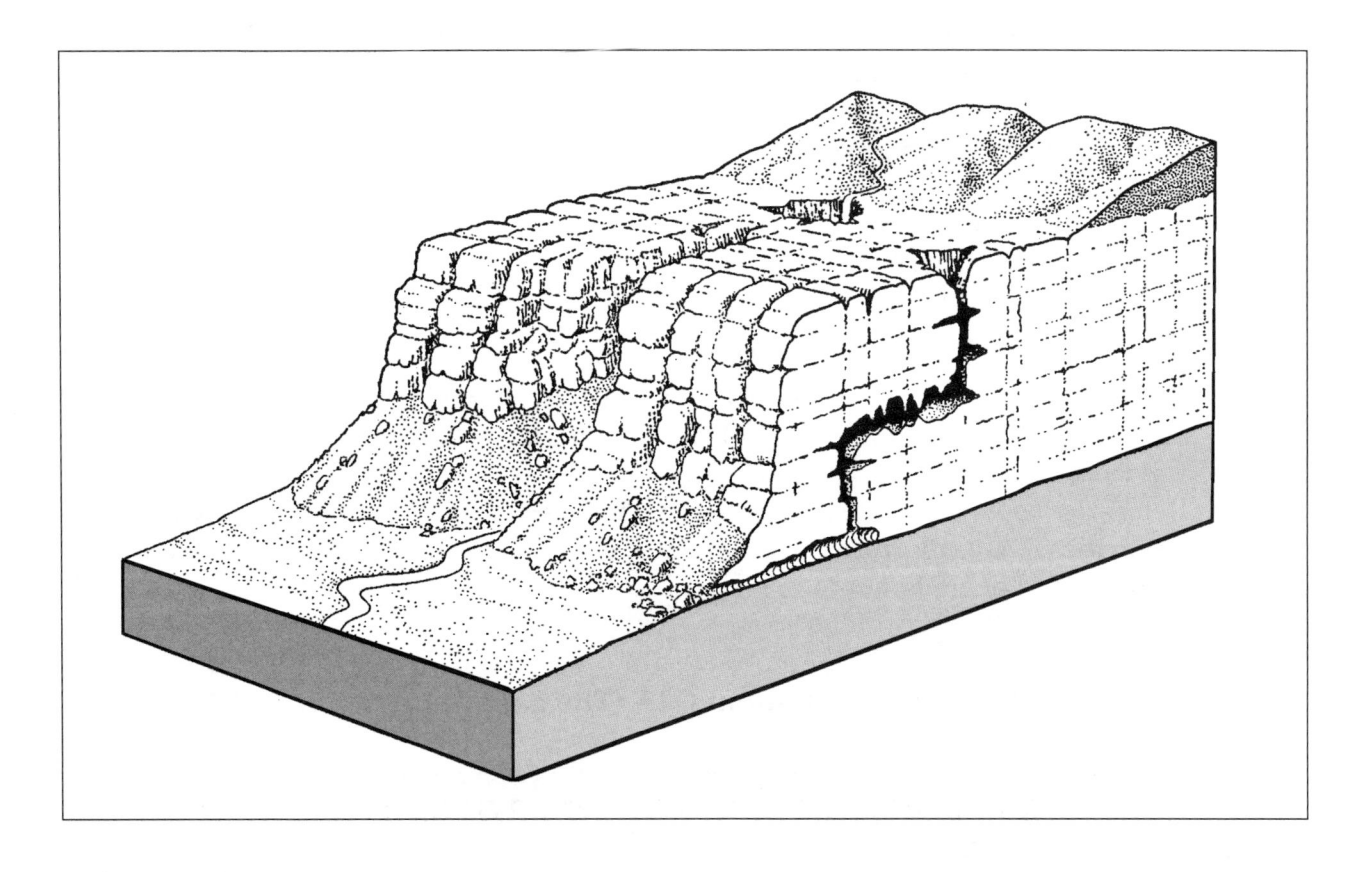

Reference Diagram Q1B (Conservation Strategies)

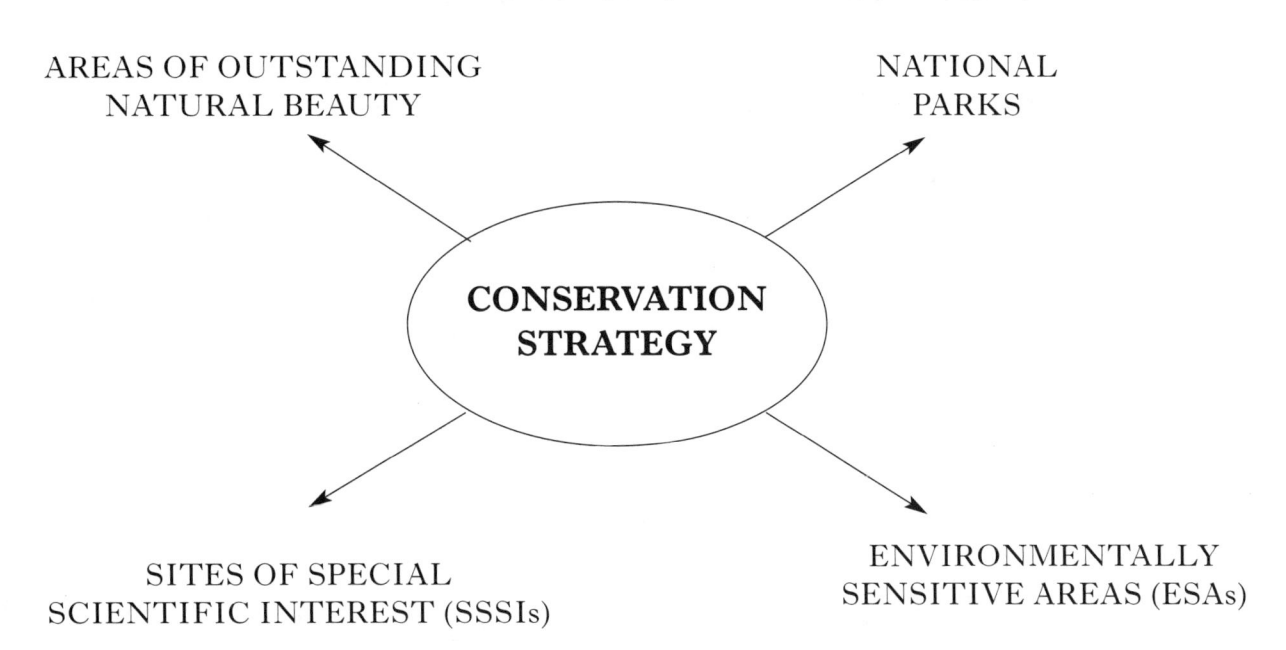

AREAS OF OUTSTANDING NATURAL BEAUTY

NATIONAL PARKS

CONSERVATION STRATEGY

SITES OF SPECIAL SCIENTIFIC INTEREST (SSSIs)

ENVIRONMENTALLY SENSITIVE AREAS (ESAs)

[Turn over

Question 2 (Rural Land Degradation)

(a) Study Reference Map Q2A and Reference Maps Q2B.

Describe the climatic conditions found in Burkina Faso, and **explain** why such conditions may lead to the degradation of rural land. **16**

(b) **Explain** how inappropriate farming activities such as overcultivation, monoculture, overgrazing, poor irrigation techniques and inappropriate cultivation of marginal land have led to land degradation in some named areas of North America. **8**

(c) Study Reference Statements Q2A and Q2B.

Select **one** of the statements and **explain** how degradation has impacted on the social and economic ways of life in that area. **10**

(d) Referring to named areas in North America that you have studied, **describe** and **explain** ways in which changes in farming methods have reduced land degradation. **16**

(50)

Reference Statement Q2A (The Sahel)

28·5 MILLION PEOPLE ARE AFFECTED BY DESERTIFICATION IN THE SAHEL REGION OF AFRICA.

Reference Statement Q2B (The Amazon Basin)

179 000 SQUARE KILOMETRES OF RAINFOREST HAVE DISAPPEARED IN THE AMAZON BASIN SINCE 1997, EQUIVALENT TO 78% OF THE TOTAL AREA OF BRITAIN.

Question 2 — continued

Reference Map Q2A (Climatic Regions of Burkina Faso)

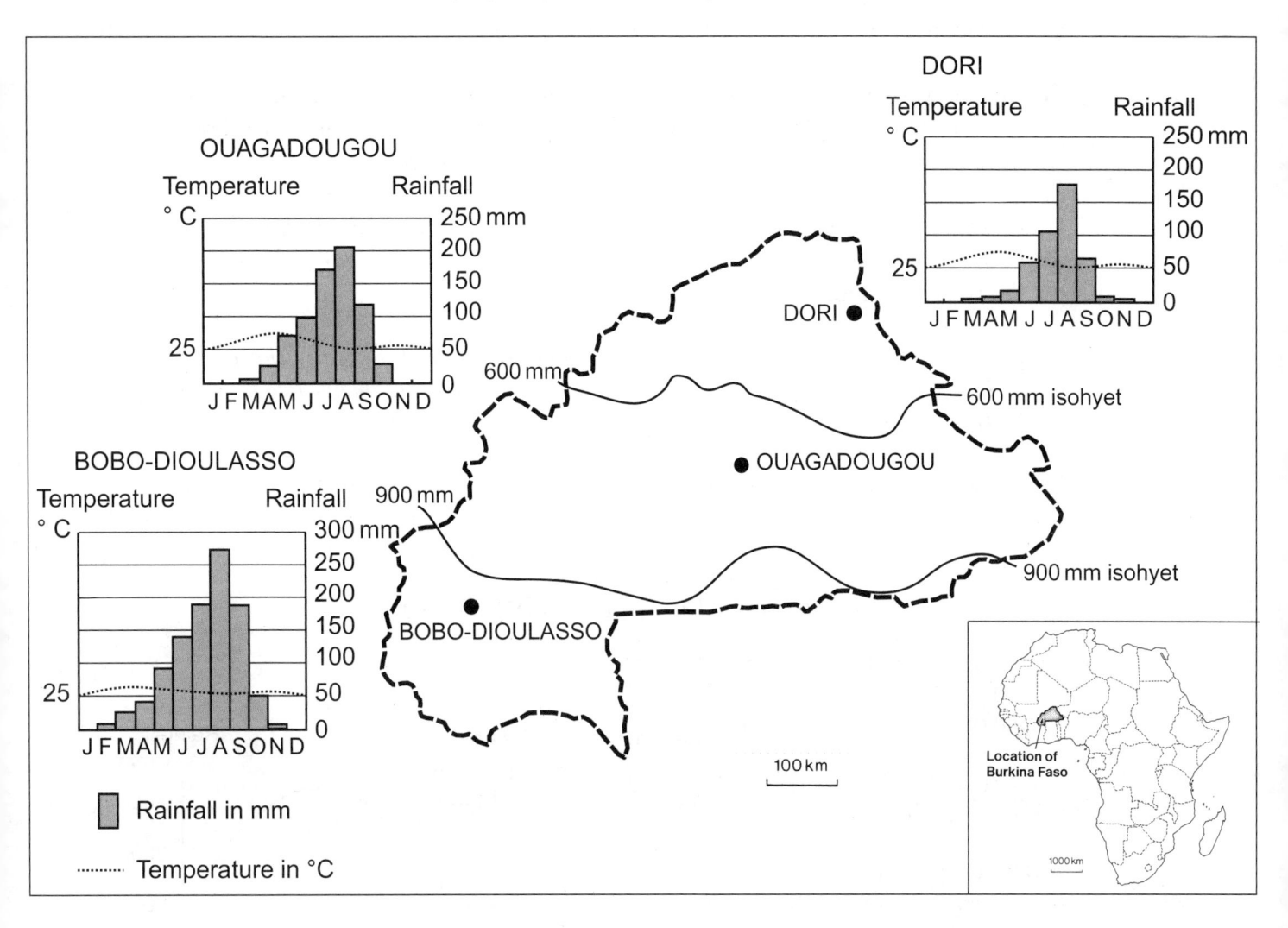

Reference Maps Q2B (Burkina Faso: mean annual rainfall patterns)

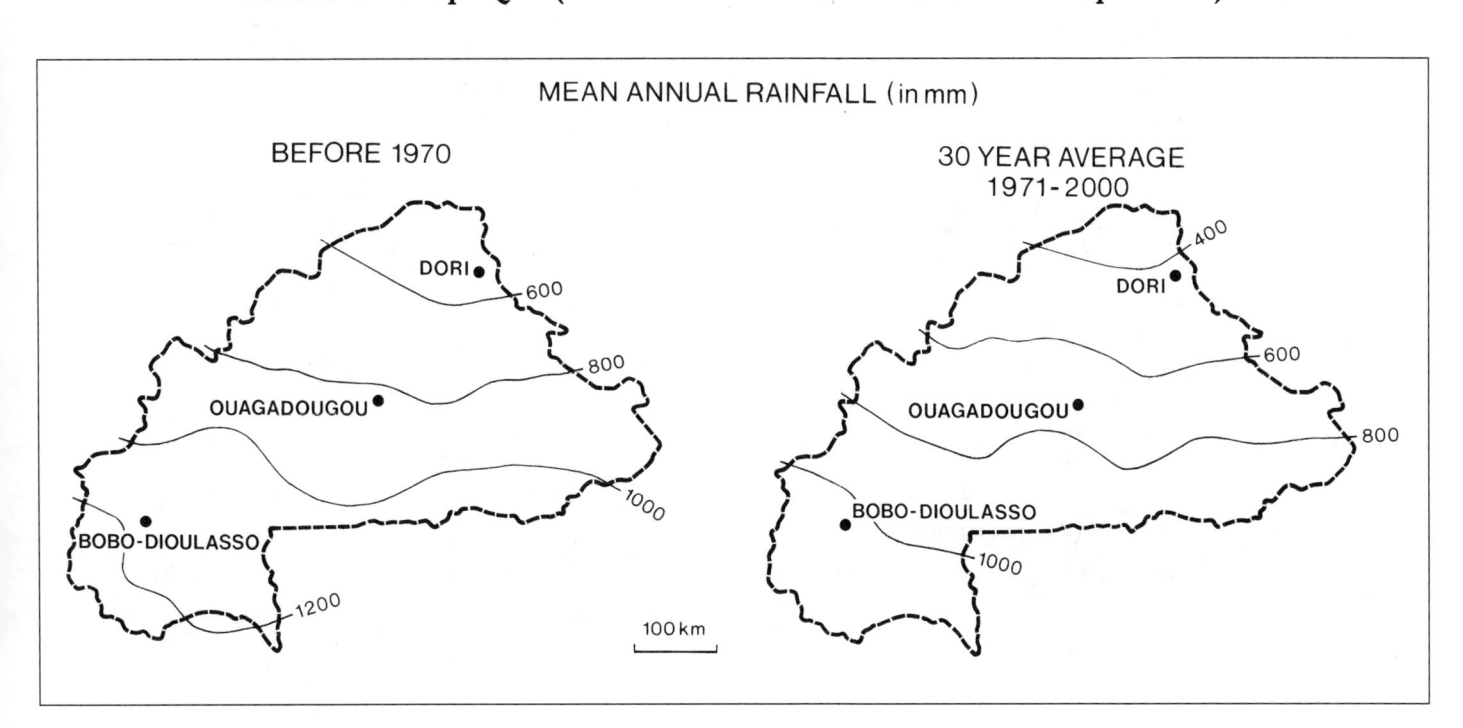

Question 3 (River Basin Management)

(a) Study Reference Map Q3 and Reference Diagram Q3.

   Explain why there is a need for water management in Egypt. **10**

(b) For the Aswan High Dam **or** any dam you have studied in Africa **or** North America **or** Asia, **explain** the **physical** factors which should be considered when selecting the site for the dam and associated reservoir. **10**

(c) **Describe** and **explain** the social, economic and environmental benefits **and** adverse consequences of a named major water control project in Africa **or** North America **or** Asia. **24**

(d) *"Potential 'water wars' are likely in areas where rivers and lakes are shared by more than one country or state, according to a UN Development Programme (UNDP) report."*

   Explain why political problems can occur in the development of water control projects. **6**

   (50)

Reference Map Q3 (The Nile Basin)

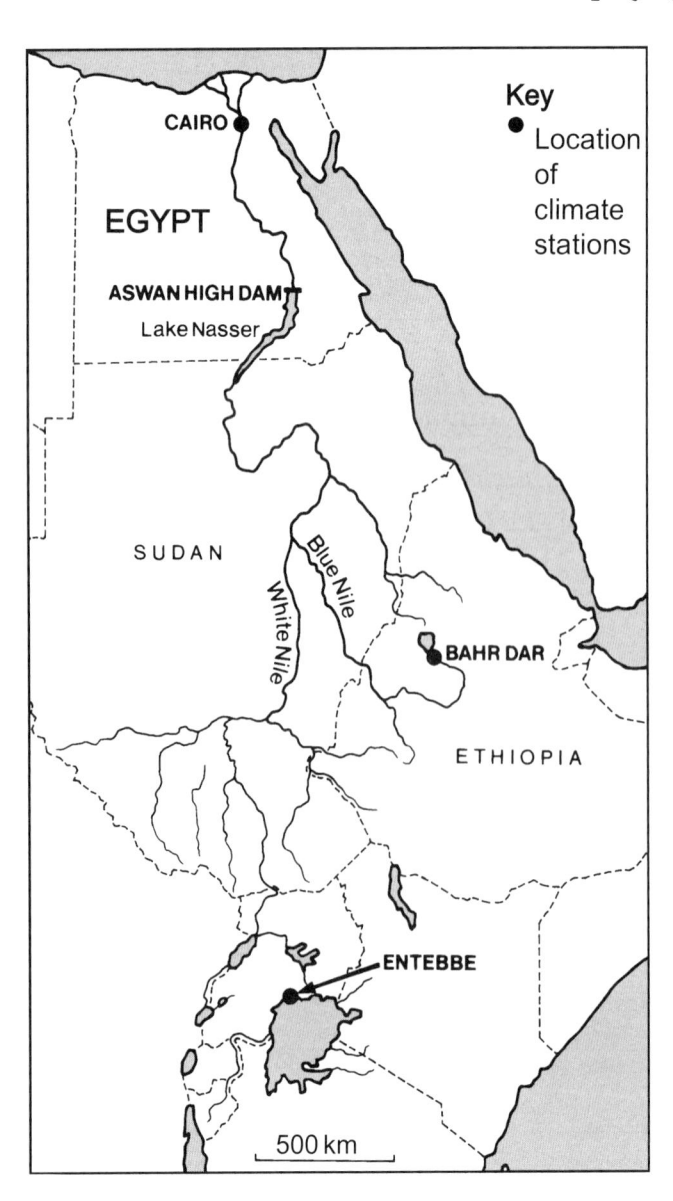

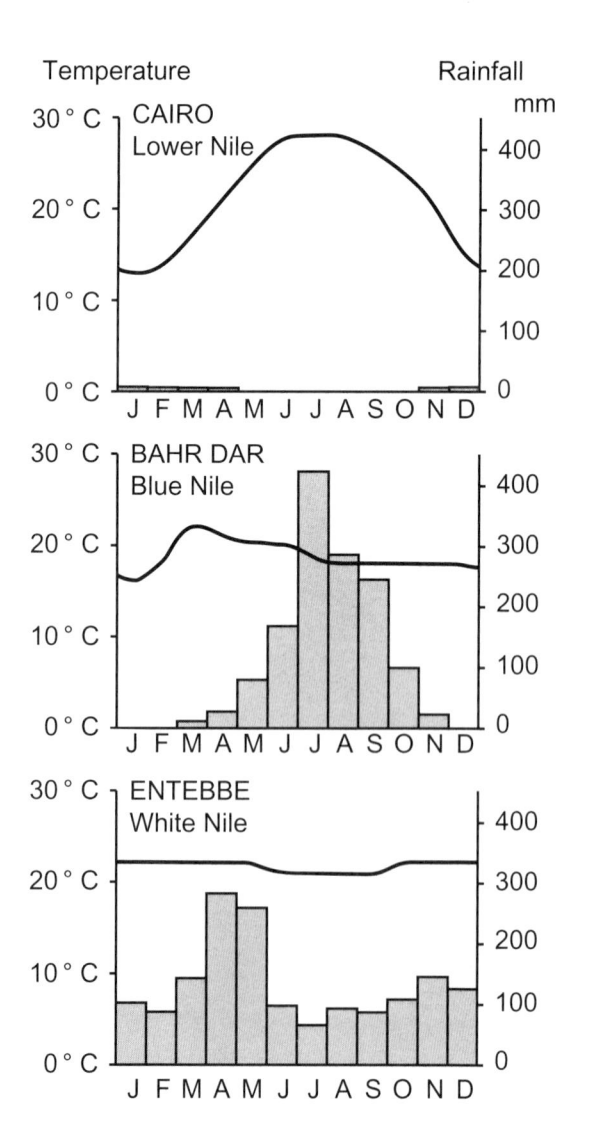

Question 3 — continued

Reference Diagram Q3 (Population of Egypt (1950–2050)

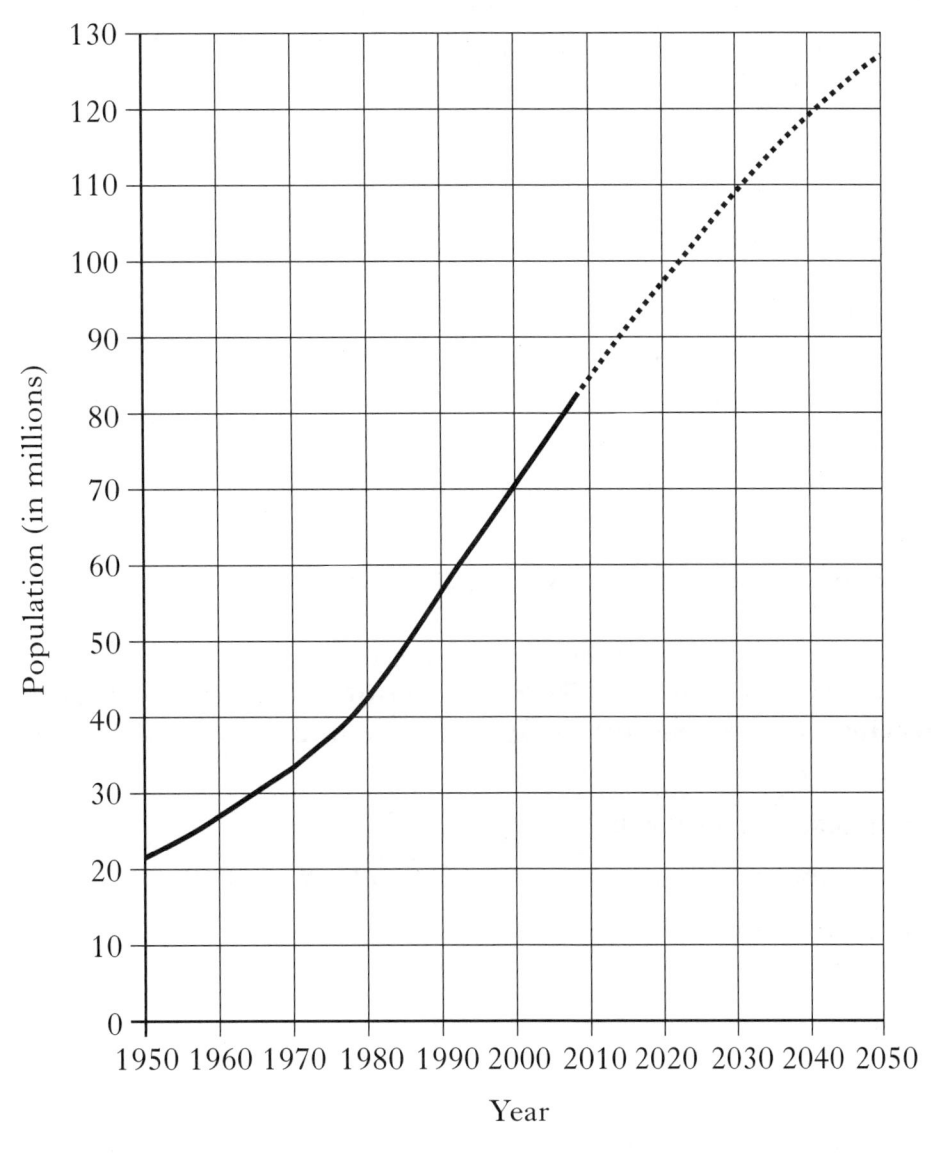

— actual figures

······ estimate

[Turn over

SECTION 2

You must answer ONE question from this Section.

Question 4 (Urban Change and its Management)

(a) Study Reference Table Q4A.

  (i) **Describe** the changes shown in the table.

  (ii) **Suggest** reasons for the differences between more developed and less developed regions. **14**

(b) With the aid of Reference Photograph Q4 and referring to a named city which you have studied in an **ELDC (Economically Less Developed Country)**:

  (i) **describe** the social, economic and environmental problems created by shanty towns; and

  (ii) **describe ways** in which such problems are being tackled. **18**

(c) Study Reference Table Q4B which highlights problems which have occurred in cities in **EMDCs (Economically More Developed Countries)** over the last fifty years. Choose **one** of these problems and, with reference to a named city in an EMDC:

  (i) **suggest** reasons for the problem;

  (ii) **describe** strategies used to solve the problem; and

  (iii) **comment** on the success of these strategies. **18**

(50)

Question 4 — continued

Reference Table Q4A (Percentage of total population living in urban areas)

Region	Urban population (%)		
	1970	1994	2025
More developed regions	**67**	**75**	**84**
Europe	64	73	83
North America	74	76	85
Less developed regions	**25**	**37**	**57**
Africa	23	33	54
South and Central America	57	74	85

Reference Photograph Q4 (A Shanty Town in Cape Town, South Africa)

Reference Table Q4B (Selected problems facing cities in EMDCs (Economically More Developed Countries))

- Housing change in the Inner City
- The decline of traditional industries
- The rise of out of town shopping

Marks

Question 5 (European Regional Inequalities)

(a) Study Reference Map Q5.

The European Union (EU) is often said to fit the "Core and Periphery" model. Ten of the twelve countries which joined the EU since 2004 have formed a new "Eastern Periphery".

Suggest both physical **and** human reasons for the lack of prosperity in the new "Eastern Periphery". **10**

(b) Study Reference Table Q5 which shows a range of indicators for six European Union countries.

Describe and **explain** the ways in which the information shows the **differences** between the three **groups** of countries shown in the table. **12**

(c) "There are marked differences in economic development within the United Kingdom (UK)."

Describe and **explain** both the physical **and** human factors that have led to regional inequalities within the UK. **14**

(d) For **either** the UK **or** another named country in the EU which has marked regional differences in economic development, **discuss** ways in which the National Government **and** the EU have tried to tackle problems in less prosperous regions. **14**

(50)

Question 5 — continued

Reference Map Q5 (The Core and Eastern Periphery of the European Union)

Reference Table Q5 (Selected indicators of development)

Economic group	Country	Infant mortality rate (per 1000 live births)	GDP per capita ($)	Employment (%) Primary	Employment (%) Secondary	Employment (%) Tertiary
Euro-Core Countries	Germany	4	30 400	2	26	72
Euro-Core Countries	Netherlands	5	30 500	2	19	79
Pre-2004 Periphery Countries	Greece	5	22 200	12	20	68
Pre-2004 Periphery Countries	Portugal	5	19 300	10	30	60
Eastern Periphery Countries	Estonia	8	16 700	11	20	69
Eastern Periphery Countries	Slovakia	7	16 000	6	29	65

Marks

Question 6 (Development and Health)

(a) Study Reference Map Q6.

   (i) **Describe** clearly **two** economic and **two** social indicators of development which could be used to produce a map such as this. **8**

   (ii) **Suggest reasons** for the wide variations in development which exist **between** Economically Less Developed Countries (ELDCs).

   You should refer to named ELDCs you have studied. **12**

   (iii) There are often considerable differences in levels of development and living standards **within** a single country.

   Referring to a named ELDC which you have studied, **suggest reasons** why such regional variations exist. **10**

(b) For **either** malaria **or** bilharzia **or** cholera:

   (i) **describe** the environmental **and** human factors which put people at risk of contracting the disease; and **8**

   (ii) **describe** and **evaluate** the methods used to control the spread of the disease. **12**

   (50)

Question 6 — continued

Reference Map Q6 (The World: Human Development Index (HDI))

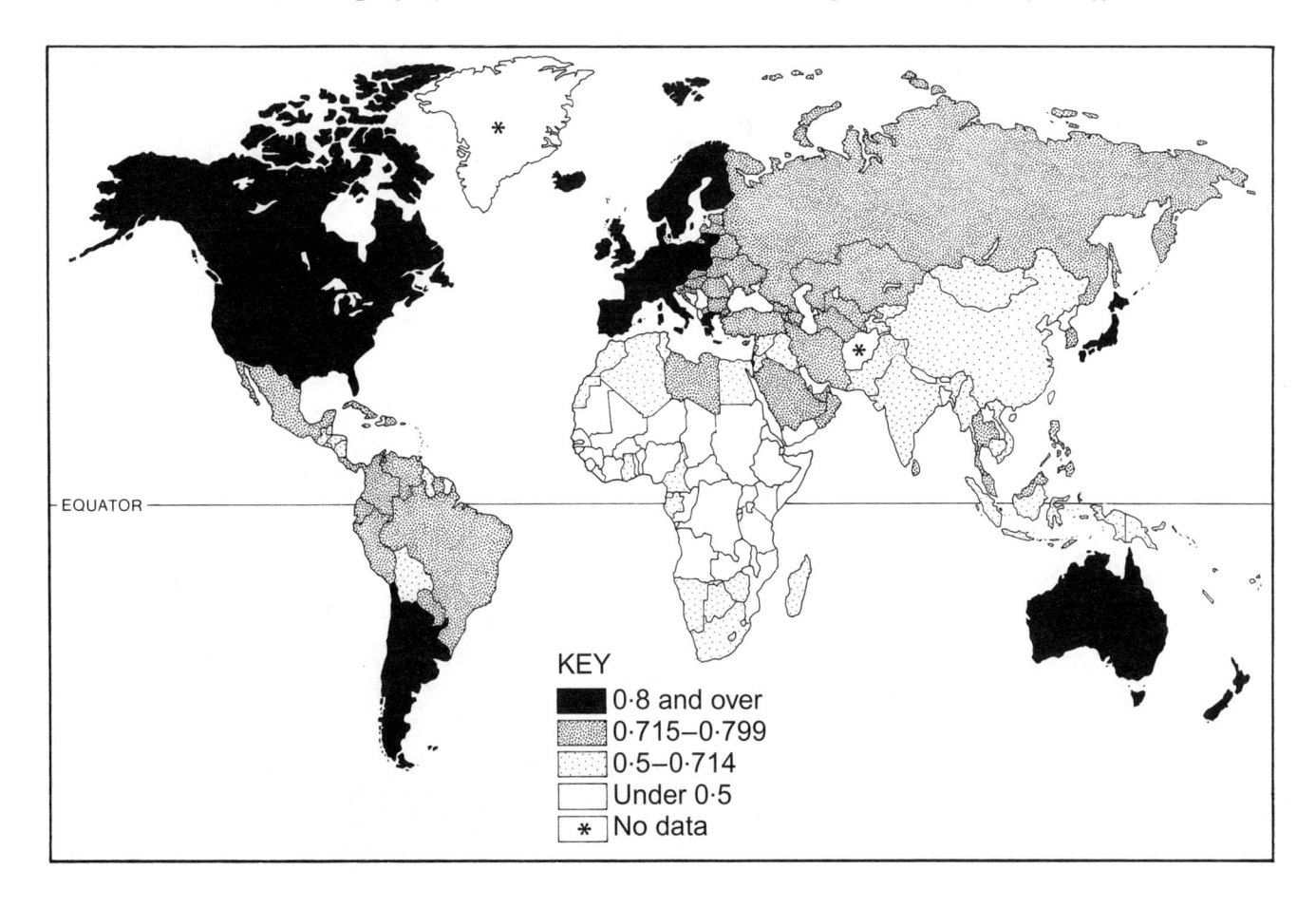

KEY
- 0·8 and over
- 0·715–0·799
- 0·5–0·714
- Under 0·5
- * No data

[END OF QUESTION PAPER]

HIGHER
2009

[BLANK PAGE]

X208/301

NATIONAL
QUALIFICATIONS
2009

WEDNESDAY, 27 MAY
9.00 AM – 10.30 AM

GEOGRAPHY
HIGHER
Paper 1
Physical and
Human Environments

Six questions should be attempted, namely:

all four questions in **Section A** (Questions 1, 2, 3 and 4);

one question from **Section B** (Question 5 **or** Question 6);

one question from **Section C** (Question 7 **or** Question 8).

Write the numbers of the **six** questions you have attempted in the marks grid on the back cover of your answer booklet.

The value attached to each question is shown in the margin.

Credit will be given for appropriate maps and diagrams, and for reference to named examples.

Questions should be answered in sentences.

Note The reference maps and diagrams in this paper have been printed in black only: no other colours have been used.

1:50 000 Scale
Landranger Series

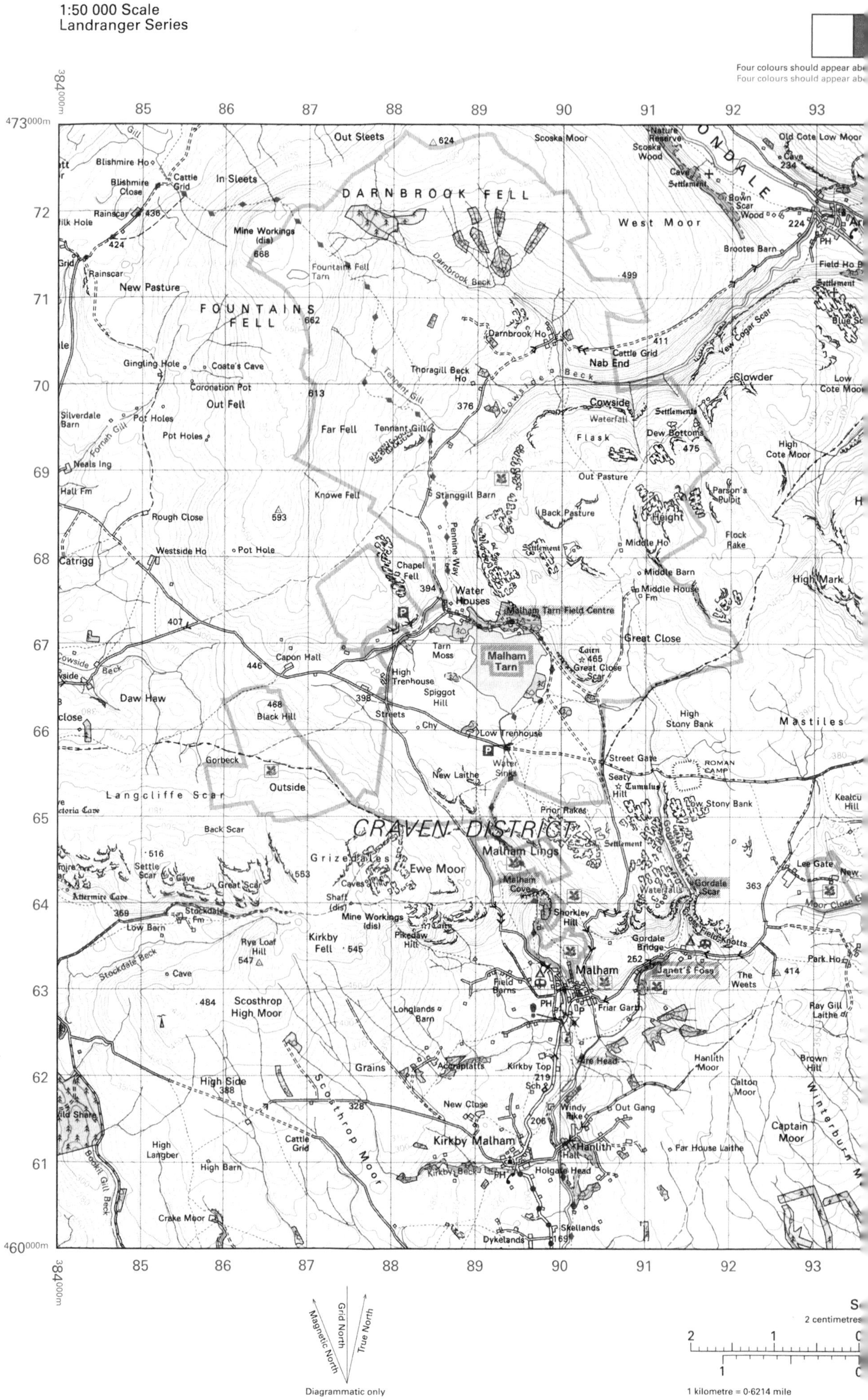

Extract No 1745/98

please return to the invigilator.
please return to the invigilator.

SECTION A: Answer ALL questions in this section

Question 1: Hydrosphere

Study OS Map Extract number 1745/98: Upper Wharfedale (*separate item*).

(a) Using appropriate grid references, **describe** the physical characteristics of the River Wharfe and its valley from 978690 to 040603.

10

(b) **Explain**, with the aid of a diagram or diagrams, how a waterfall is formed in the upper course of a river valley.

8

Question 2: Biosphere

(a) **Draw** and **fully annotate** a soil profile of a **podzol** to show its main characteristics (including horizons, colour, texture and drainage) and associated vegetation. **9**

Study Reference Diagram Q2 which shows a soil profile of a brown earth soil.

(b) **Describe** and **explain** the formation and characteristics of a **brown earth soil**. **9**

Reference Diagram Q2

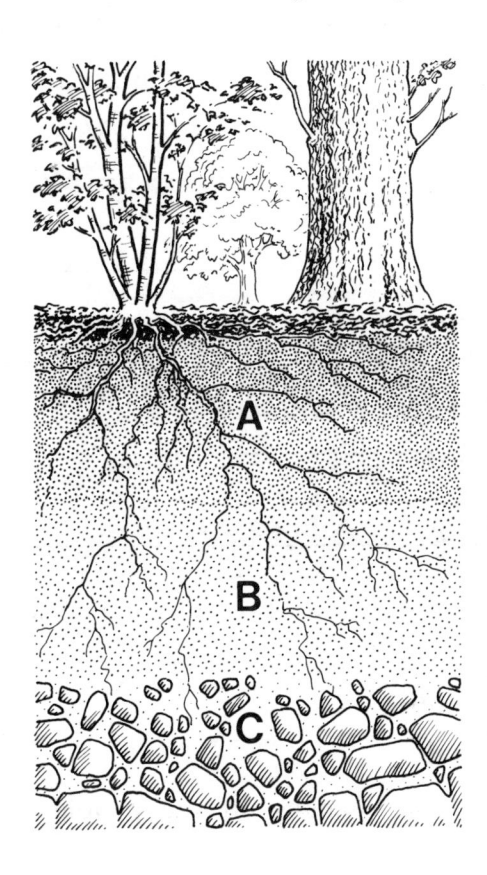

Marks

Question 3: Rural Geography

(a) Study Reference Table Q3.

Describe and **explain**, with the aid of the data in the table, the differences between intensive peasant farming and commercial arable farming.

10

Reference Table Q3 (Types of farming and selected data)

	Bangladesh	Canada
Farm type	Intensive peasant farming	Commercial arable farming
GDP per capita ($US)	158	13 034
% GDP from farming	48	3
% population engaged in farming	82	4
Kg of fertiliser used per hectare	26	32
People per tractor	20 581	38

(b) Areas of intensive peasant farming such as those in Bangladesh have undergone changes in recent years.

Referring to an area you have studied:

(i) **describe** these changes, and

(ii) **outline** the impact of these changes on the people **and** the farming landscape.

10

Marks

Question 4: Industrial Geography

"*Many industrial concentrations within the European Union have undergone a great transformation in the last 50 years. These changes are most marked in the types of industries, the industrial landscape and in employment patterns.*"

Referring to a **named** industrial concentration in the European Union that you have studied:

(i) **describe** and **account for** the main characteristics of a typical "new" industrial landscape; **9**

(ii) **suggest** ways in which the national government **and** the European Union have helped to attract new industries to your chosen area. **7**

[Turn over

Marks

SECTION B: Answer ONE question from this section, ie either Question 5 or Question 6.

Question 5: Atmosphere

"Energy is transferred from areas of surplus, between 35°N and 35°S, to areas of deficit, polewards from 35°N and 35°S, by both oceanic and atmospheric circulation."

Study Reference Map Q5 which shows selected ocean currents in the North Atlantic Ocean.

(a) (i) **Describe** the pattern of ocean currents in the North Atlantic Ocean, and

 (ii) **explain** how they help to maintain the global energy balance.   6

Reference Map Q5 (Selected ocean currents in the North Atlantic Ocean)

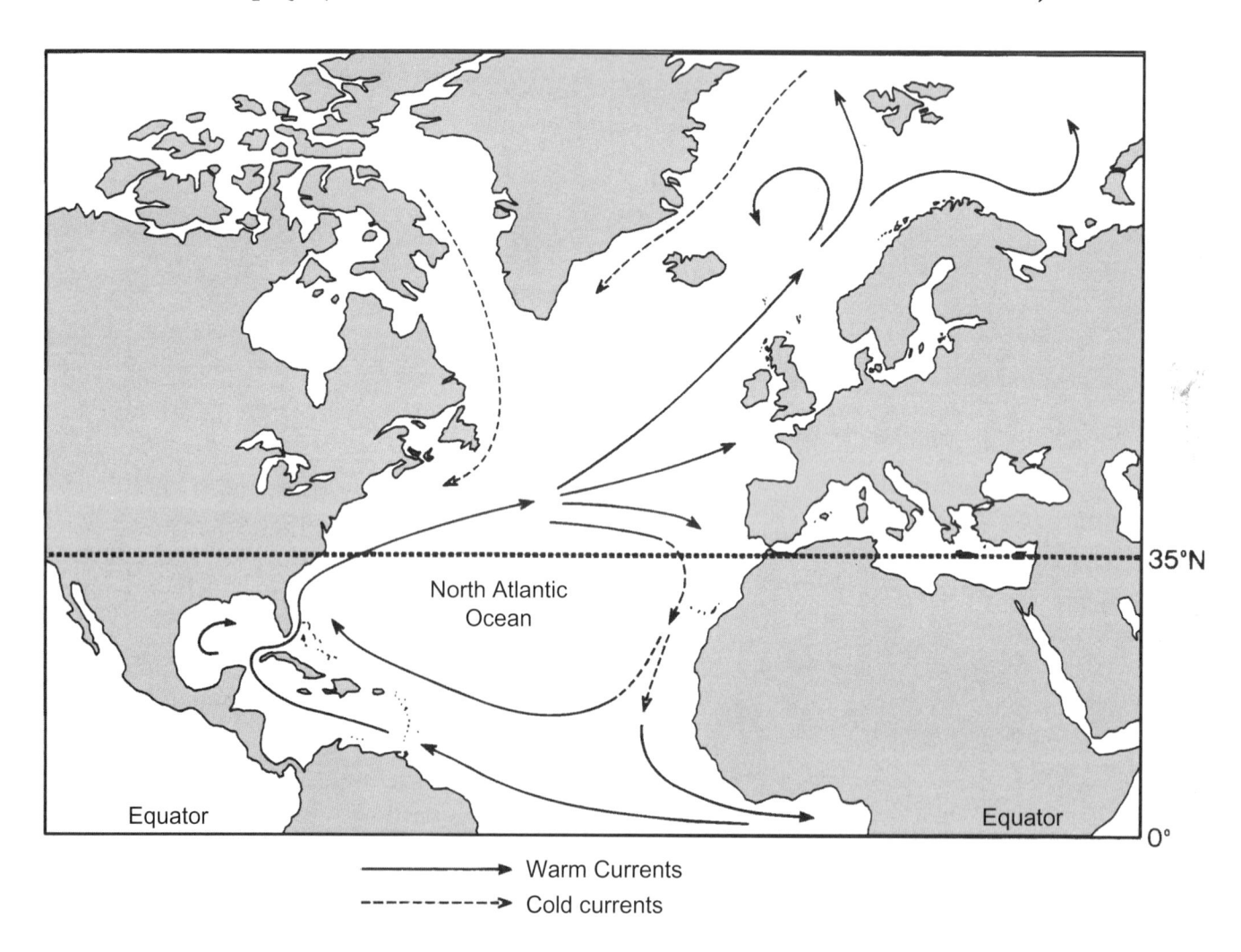

Marks

Question 5: Atmosphere (continued)

Study Reference Diagram Q5 which shows surface winds and pressure zones.

(b) **Explain** how circulation cells in the atmosphere and the associated surface winds assist in the transfer of energy between areas of surplus and deficit.

8

Reference Diagram Q5 (Surface winds and pressure zones)

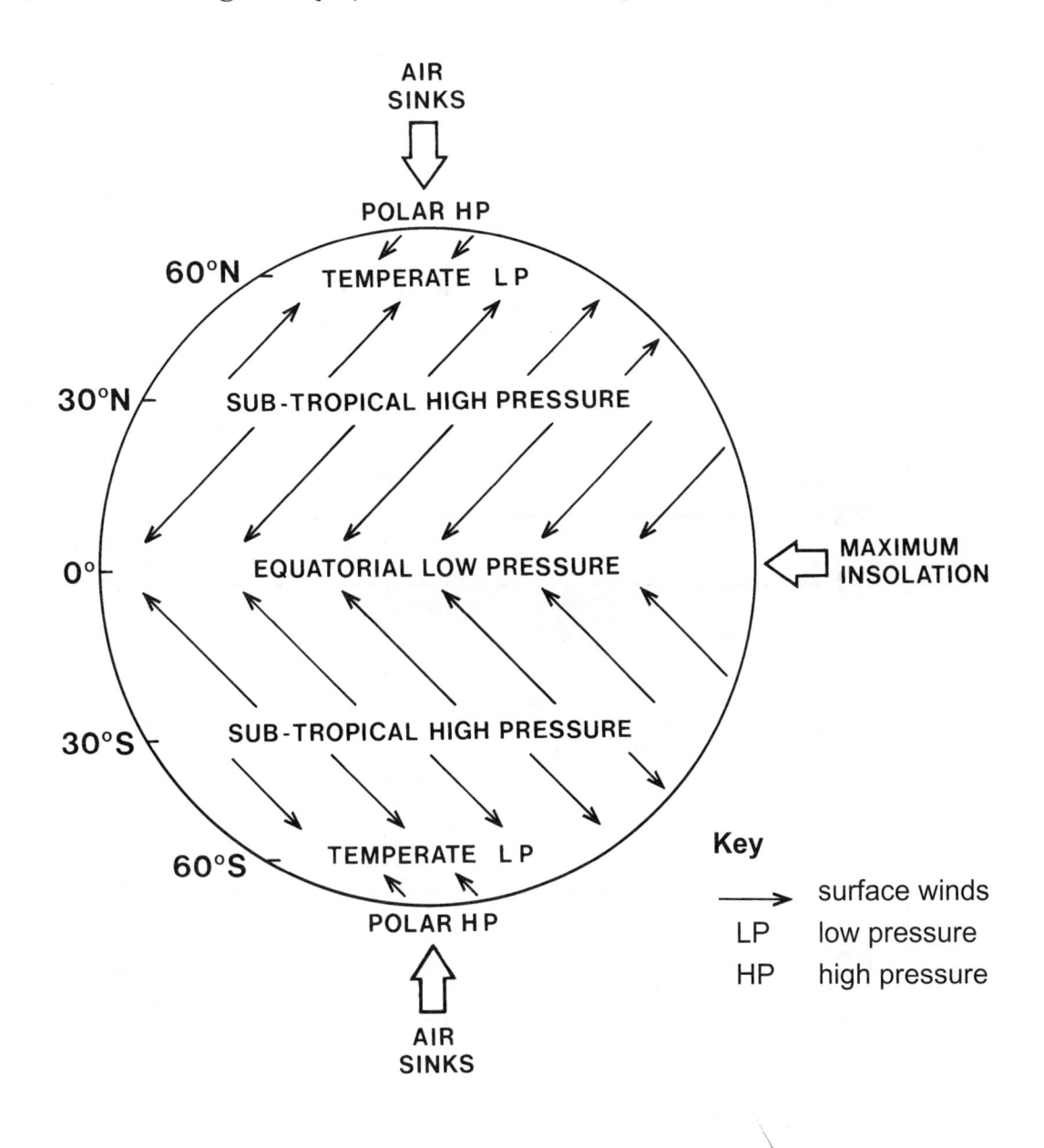

[Turn over

DO NOT ANSWER THIS QUESTION IF YOU HAVE ALREADY ANSWERED QUESTION 5

Marks

Question 6: Lithosphere

Study OS Map Extract number 1745/98: Upper Wharfedale (*separate item*), and Reference Map Q6.

The map extract covers part of the Yorkshire Dales National Park, an area famed for its Carboniferous Limestone scenery, characterised by distinctive surface features, drainage patterns and underground landforms.

(*a*) **Describe** the evidence which suggests that Area A, shown on Reference Map Q6, is a Carboniferous Limestone landscape.
(You should refer to named features and make use of grid references.) 8

(*b*) Choose any **one** Carboniferous Limestone feature described in your answer to part (*a*) and, with the aid of annotated diagrams, **explain** how it was formed. 6

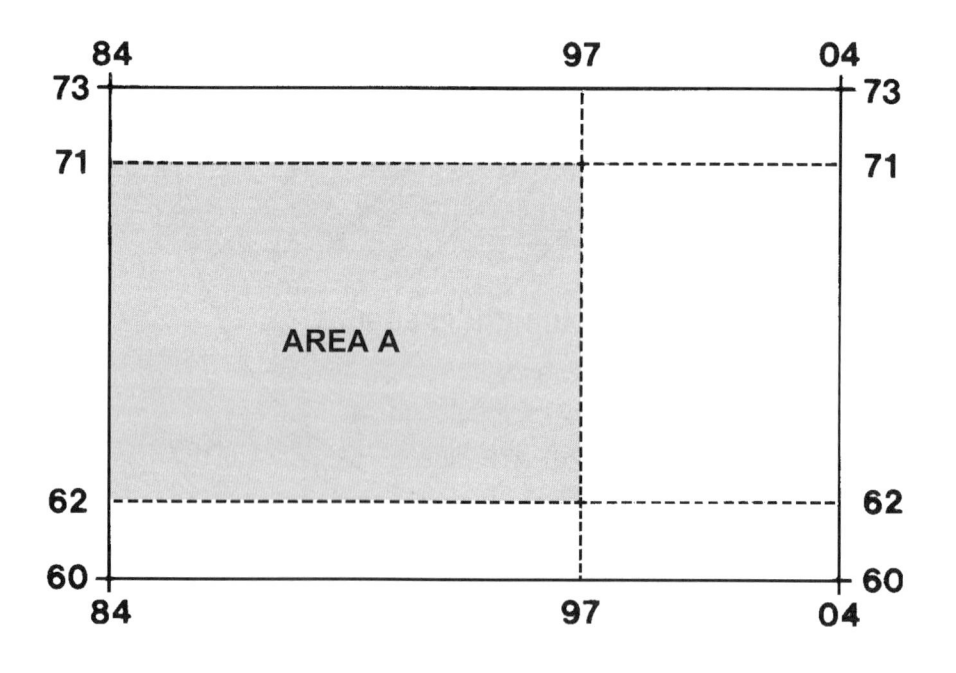

Reference Map Q6

Marks

SECTION C: Answer ONE question from this section,
ie either Question 7 or Question 8.

Question 7: Population

Italy has a population structure that is typical of many EMDCs (Economically More Developed Countries).

Study Reference Diagrams Q7A and Q7B.

(a) **Describe** and **account** for the changes between the population structure in 2000 and that projected for 2050.

8

(b) **Discuss** the consequences of the 2050 population structure for the future economy of the country and the welfare of its citizens.

6

Reference Diagram Q7A (Italy: Population pyramid for 2000)

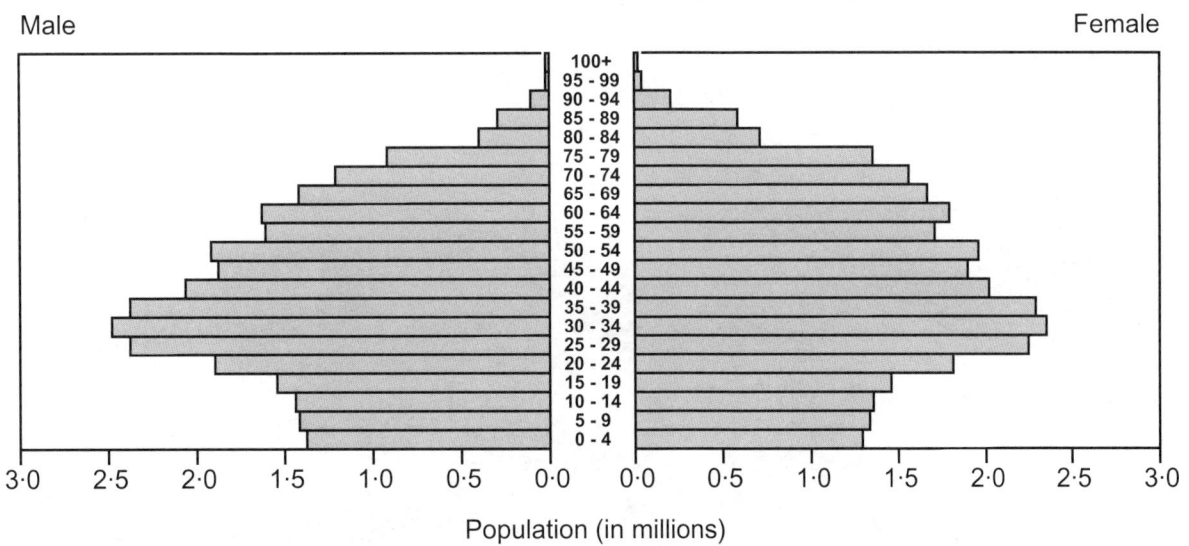

Reference Diagram Q7B (Italy: Population pyramid for 2050)

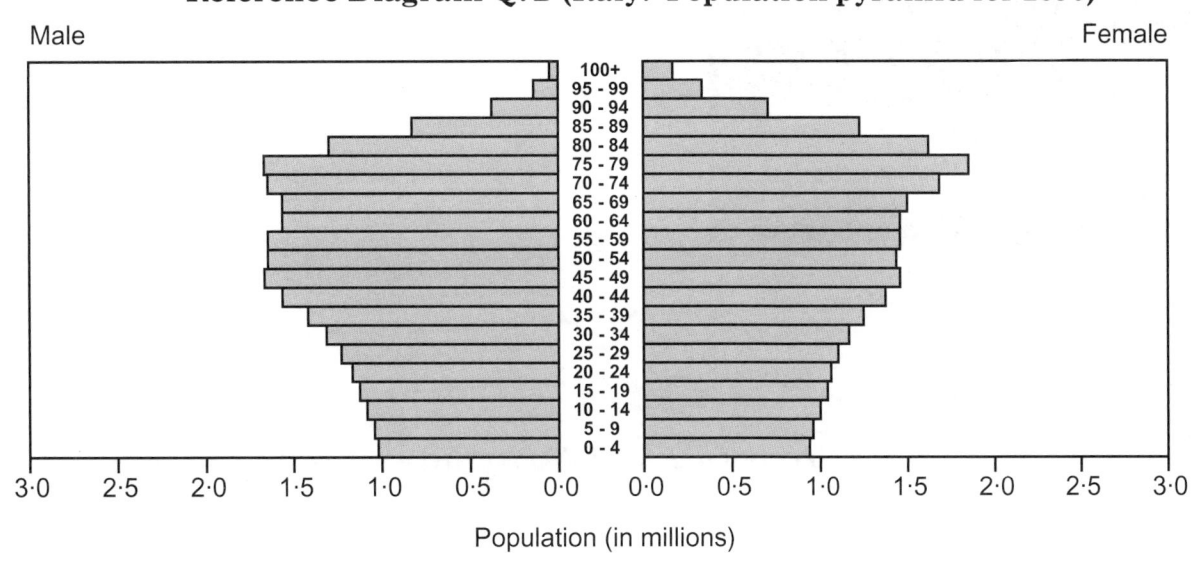

[Turn over

DO NOT ANSWER THIS QUESTION IF YOU HAVE ALREADY ANSWERED QUESTION 7

Question 8: Urban Geography

Study Reference Photograph Q8A which shows Buchanan Galleries shopping centre in Glasgow's CBD and Reference Photograph Q8B which shows Braehead, an out-of-town shopping centre situated at the south-west edge of Glasgow.

Referring to Glasgow, **or** any other named city you have studied in an Economically More Developed Country (EMDC):

(i) **suggest** the impact that an out-of-town shopping centre may have had on shopping in the traditional CBD; 6

(ii) **describe** and **explain** the changes, other than shopping, which have taken place in the CBD over the past few decades. 8

Reference Photograph Q8A

Reference Photograph Q8B

[*END OF QUESTION PAPER*]

X208/303

NATIONAL
QUALIFICATIONS
2009

WEDNESDAY, 27 MAY
10.50 AM – 12.05 PM

GEOGRAPHY
HIGHER
Paper 2
Environmental
Interactions

Answer any **two** questions.

Write the numbers of the **two** questions you have attempted in the marks grid on the back cover of your answer booklet.

The value attached to each question is shown in the margin.

Credit will be given for appropriate maps and diagrams, and for reference to named examples.

Questions should be answered in sentences.

Note The reference maps and diagrams in this paper have been printed in black only: no other colours have been used.

Marks

Question 1 (Rural Land Resources)

Loch Lomond and the Trossachs became Scotland's first National Park in 2002. It covers 1865 square kilometres of lowland, river, loch, forest and mountain landscapes.

(a) **Describe** and **explain**, with the aid of annotated diagrams, the formation of the main glacial features of the Loch Lomond and the Trossachs National Park **or** any other glaciated upland area in the UK that you have studied. **20**

(b) With reference to Loch Lomond and the Trossachs **or** any other named upland area that you have studied, **explain** the social and economic opportunities created by the landscape. **10**

(c) Study Reference Diagram Q1.

Reference Diagram Q1 shows the Loch Lomond and the Trossachs National Park to be under intense environmental pressure in certain key areas. With reference to this area or any **named** upland area you have studied:

(i) **describe** and **explain** the environmental conflicts that may occur (you should refer to named locations within your chosen upland landscape); **10**

(ii) **describe** specific solutions to these environmental conflicts commenting on their effectiveness. **10**

(50)

Question 1 — continued

Reference Diagram Q1 (Loch Lomond and the Trossachs: Environmental Activity and Pressure)

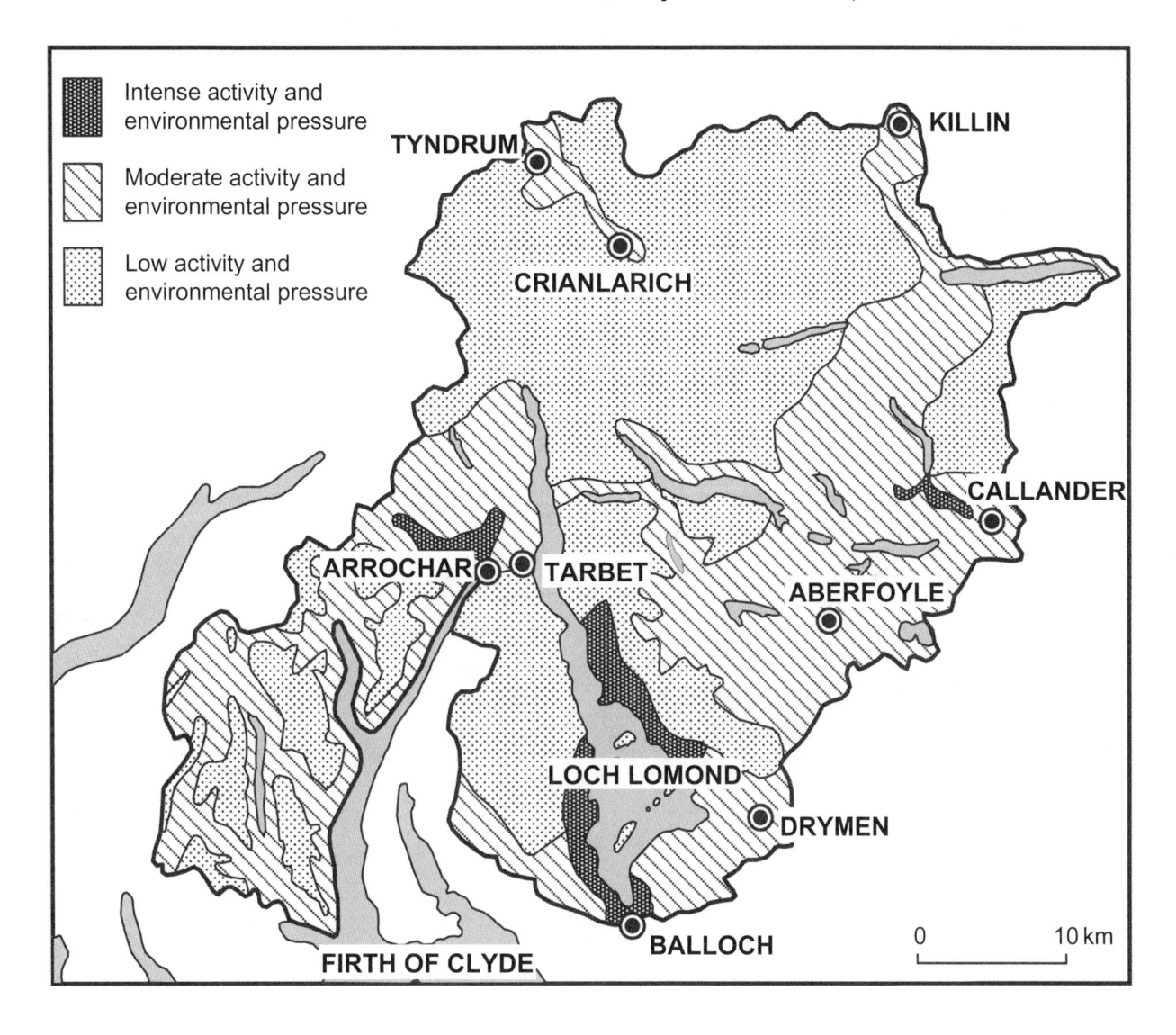

Marks

Question 2 (Rural Land Degradation)

The Sahel is a 500 kilometre wide zone which runs across Africa along the southern edge of the Sahara Desert. The Sahel is under intense pressure from human activity which, combined with climate change, has created a "spiral of desertification".

(a) Study Reference Diagram Q2.

Describe the changes in rainfall patterns shown on Reference Diagram Q2.   6

(b) For **either** Africa north of the Equator **or** the Amazon Basin:

(i) **explain** how human activities, including inappropriate farming techniques, have contributed to land degradation; and   18

(ii) **describe** some of the consequences of land degradation on the people and their environment.   10

(c) Referring to **named** areas of **North America** which you have studied:

(i) **describe** some of the measures which have been taken to conserve soil and limit land degradation; and

(ii) **comment on** the effectiveness of these measures.   16

(50)

Question 2 — continued

Reference Diagram Q2 (Rainfall Variability in the Sahel)

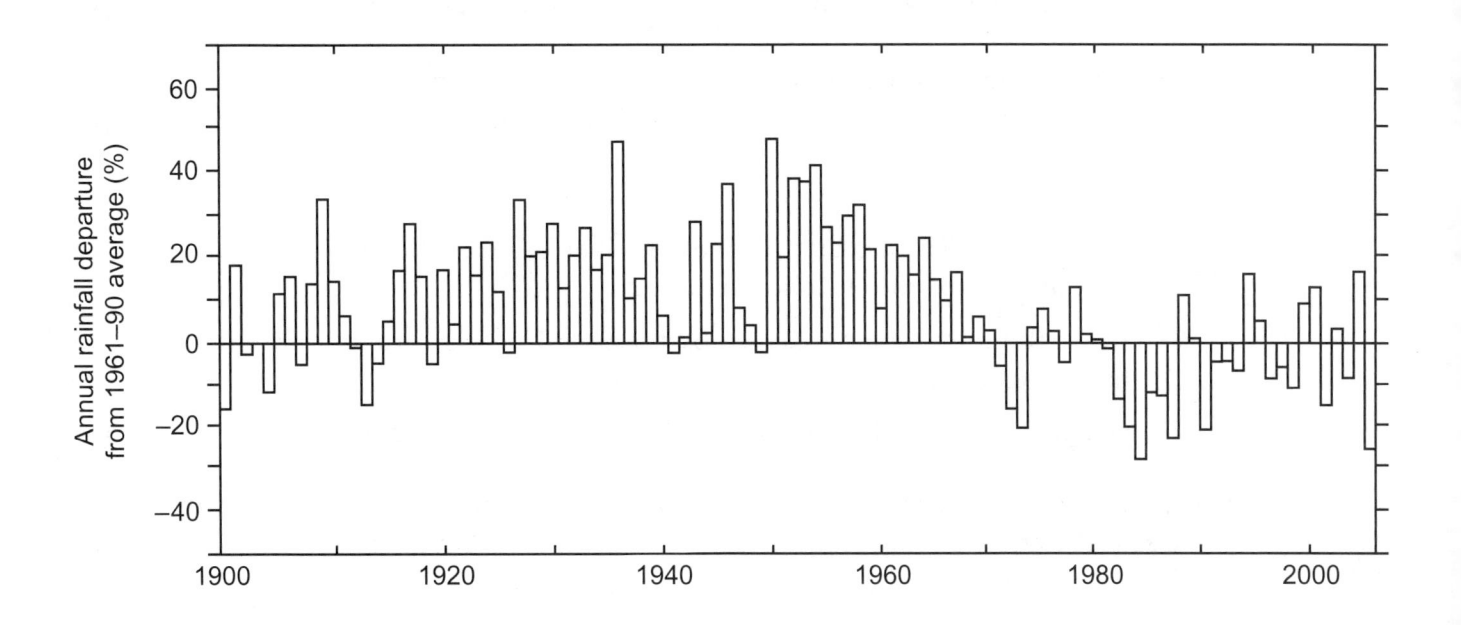

Marks

Question 3 (River Basin Management)

(a) Study Reference Table Q3 and Reference Map Q3.

Explain the need for water management in the Colorado Basin. **10**

(b) **Explain** the physical **and** human factors that have to be considered when selecting sites for dams and their associated reservoirs. **14**

(c) Study Reference Diagram Q3 and Reference Map Q3.

For the Colorado River Basin, **or** another river basin in North America, **or** in Africa, **or** in Asia, that you have studied:

(i) **describe** the problems caused by the river flowing through more than one state or country;

(ii) **suggest** ways in which these problems may be overcome. **10**

(d) **Describe** and **explain** the social, economic and environmental **benefits** of a **named** water control project in North America **or** Africa **or** Asia. **16**

(50)

Reference Table Q3 (Population Growth in Las Vegas and Phoenix)

Selected city	1990 Population	2000 Population	Population change (1990–2000)
Phoenix	2 238 480	3 251 876	+45%
Las Vegas	741 459	1 375 765	+85%

Reference Diagram Q3 (The Colorado River Water Allocation)

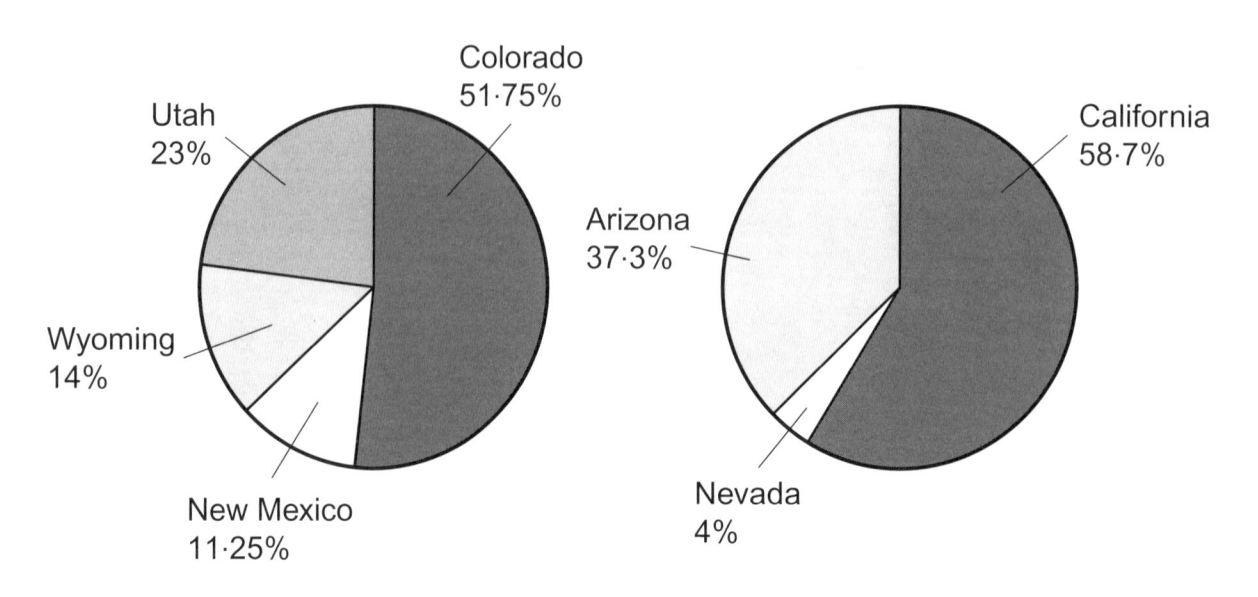

Upper Basin—Water allocation
- Colorado 51·75%
- Utah 23%
- Wyoming 14%
- New Mexico 11·25%

Lower Basin—Water allocation
- California 58·7%
- Arizona 37·3%
- Nevada 4%

Question 3 — continued

Reference Map Q3 (The Colorado River Basin)

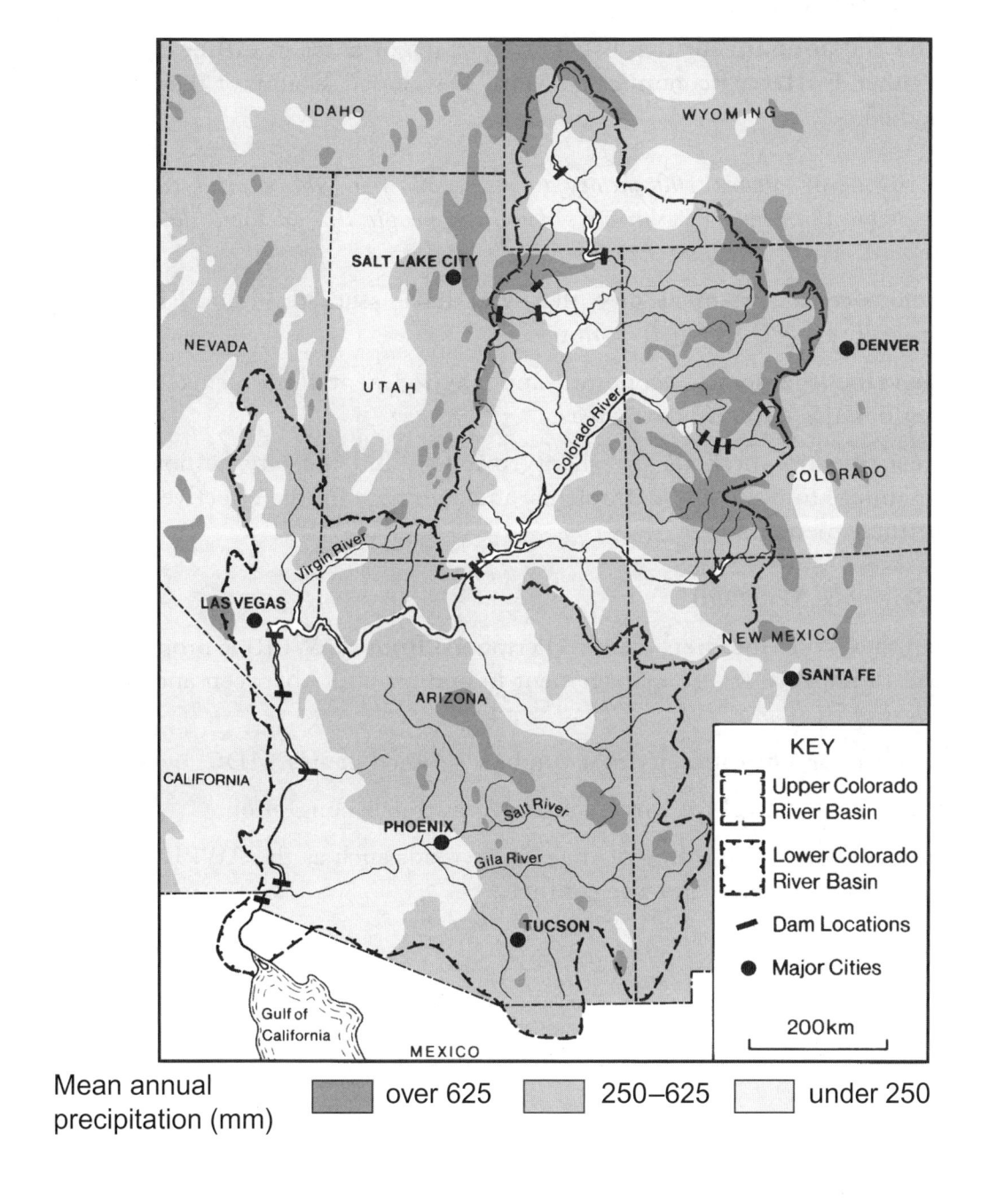

Mean annual precipitation (mm): over 625 | 250–625 | under 250

[Turn over

Marks

Question 4 (Urban Change and its Management)

(a) Study Reference Map Q4A.

Describe and **account for** the distribution of major cities in **either** Spain **or** any other EMDC (Economically More Developed Country) that you have studied. **10**

(b) "*Kibera is one of almost 100 shanty towns in Nairobi, the capital city of Kenya. More than half of Nairobi's 3 million people live in these shanties, which in total occupy less than 2% of the city's land area.*"

With reference to a named city that you have studied in an ELDC (Economically Less Developed Country):

(i) **describe** the social, economic and environmental problems often found in these shanty town areas; **12**

(ii) **describe** the methods the shanty dwellers and the city authorities might use to tackle these problems, and comment on the effectiveness of these methods. **8**

(c) Study Reference Map Q4B.

The map shows the Aberdeen Western Peripheral Route (AWPR), a proposed new road to improve traffic management in and around Aberdeen and the North-east of Scotland.

For Aberdeen, or a **named** city that you have studied in an EMDC:

(i) **describe** and **explain** why it suffers from traffic congestion; **12**

(ii) **suggest** why the building of major new roads such as the AWPR may lead to protests and land-use conflicts. **8**

(50)

Question 4 — continued

Reference Map Q4A (Largest Cities in Spain)

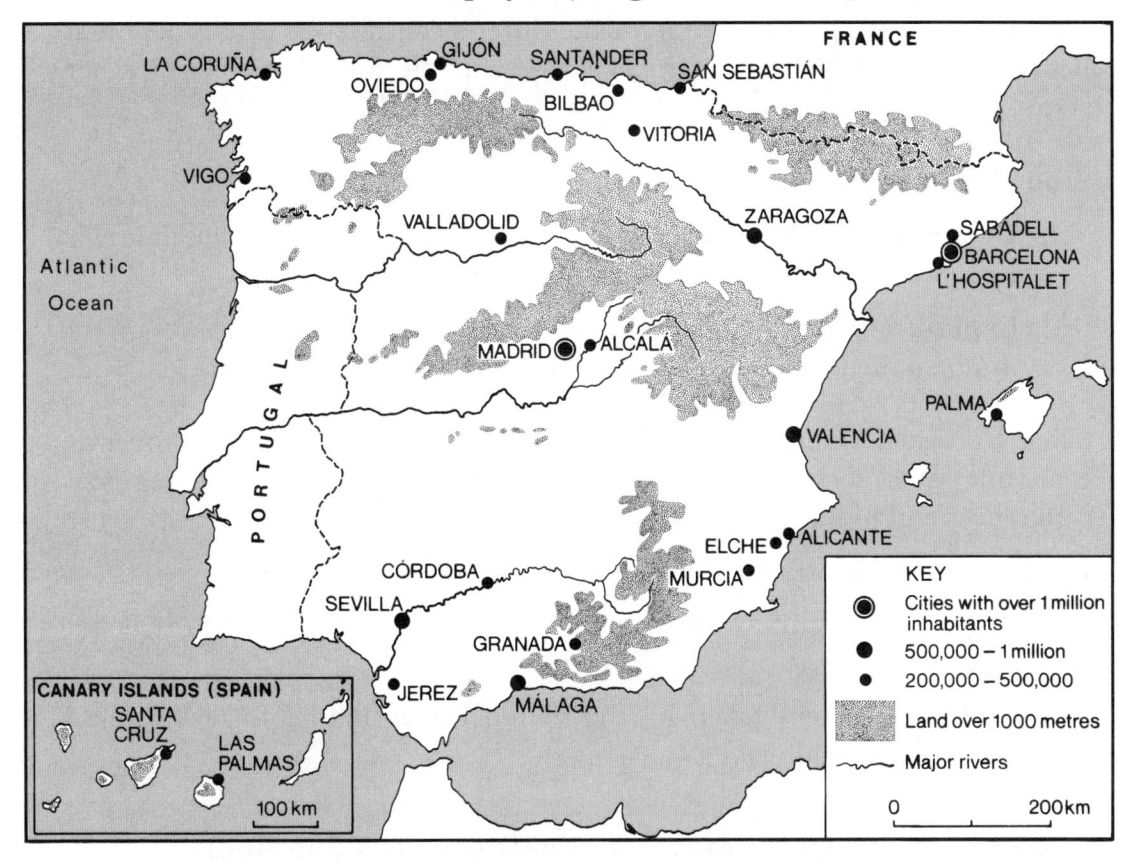

Reference Map Q4B (Aberdeen Western Peripheral Route (AWPR))

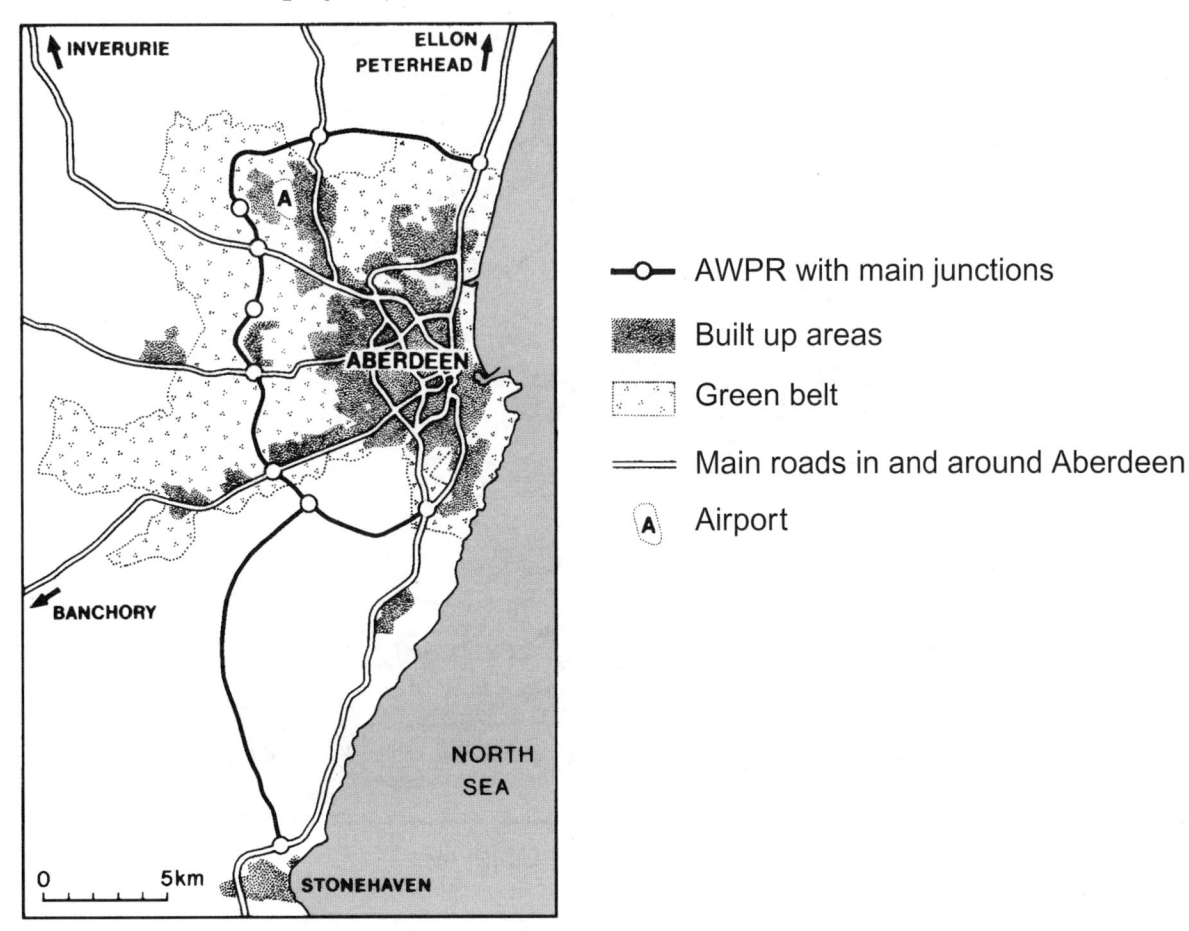

Question 5 (European Regional Inequalities)

(a) Study Reference Table Q5.

Describe and **suggest reasons** for the differences in levels of development between the pre-2000 EU member states and the post-2000 EU member states. **10**

(b) Study Reference Map Q5.

(i) **Describe** the distribution of the regions which were eligible for European grants under Objective 1 support (2000–2006). **8**

(ii) **Explain** how EU initiatives such as Objective 1 support might improve the less prosperous regions of the European Union. **8**

(c) "The European Cohesion policy (2007–2013) aims to contribute towards economic and social cohesion within the EU by reducing regional differences and human inequality within member states."

For any named country you have studied in the European Union:

(i) **describe** the physical and human factors which have led to regional inequalities; **18**

(ii) **outline** the steps taken by the national government to tackle these regional inequalities. **6**

(50)

Reference Map Q5 (European Union Objective 1 Funding)

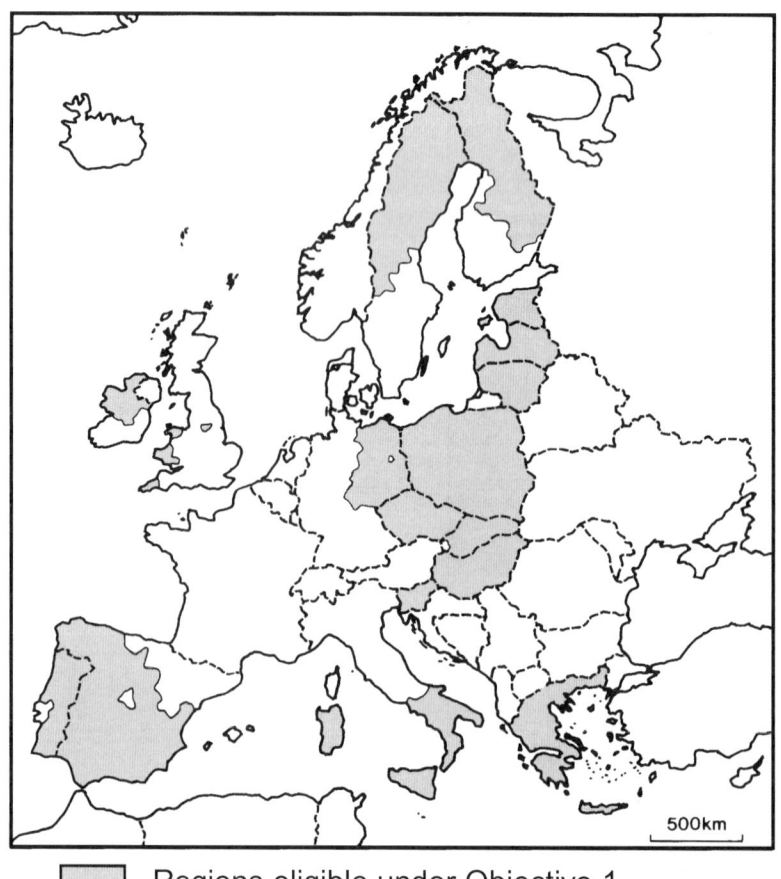

(Objective 1: Supporting development in less prosperous regions)

Question 5 — continued

Reference Table Q5 (European Union Statistics Ranked in Order)

\multicolumn{4}{c	}{Pre–2000 Member States}	\multicolumn{4}{c}{Post–2000 Member States}
Country	Year of EU membership	GDP (ranked)*	HDI (ranked)*	Country	Year of EU membership	GDP (ranked)*	HDI (ranked)*
Belgium	1957	6	6	Cyprus	2004	14	17
France	1957	11	9	Czech Rep	2004	17	18
Germany	1957	10	13	Estonia	2004	20	22
Italy	1957	12	10	Hungary	2004	21	20
Luxembourg	1957	1	5	Latvia	2004	24	25
Netherlands	1957	3	3	Lithuania	2004	23	23
Denmark	1973	5	8	Malta	2004	18	19
Ireland	1973	2	1	Poland	2004	25	21
UK	1973	8	11	Slovakia	2004	22	24
Greece	1981	15	14	Slovenia	2004	16	15
Portugal	1986	19	16	Bulgaria	2007	26	26
Spain	1986	13	12	Romania	2007	27	27
Finland	1995	9	4				
Sweden	1995	7	2				
Austria	1995	4	7				

GDP   Gross Domestic Product per capita reflects total of all goods and services per head of population

HDI   Human Development Index (covering poverty, education, health)

*Ranking   1–27 with 1 best and 27 worst

[Turn over

Marks

Question 6 (Development and Health)

(a) Study Reference Map Q6 which shows the Human Development Index (HDI) for countries of the world.

Explain the advantages of using a composite indicator of development such as the HDI rather than a single indicator. 4

(b) Referring to named examples, **suggest reasons** why there is such a wide range in levels of development **between** different ELDCs (Economically Less Developed Countries). 12

(c) For malaria, **or** bilharzia, **or** cholera:

  (i) **describe** the human and environmental factors that can contribute to the spread of the disease; 6

  (ii) **describe** the measures that have been taken to combat the disease; 12

  (iii) **explain** how the eradication or control of the disease would benefit ELDCs. 6

(d) "*Resources need to be targeted at improving Primary Health Care if we are ever going to improve the health of people in ELDCs.*" Aid worker

Describe some of the strategies involved in Primary Health Care and **explain** why these strategies for improving health standards are suited to people living in ELDCs. 10

(50)

Question 6 — continued

Reference Map Q6 (The World: Human Development Index)

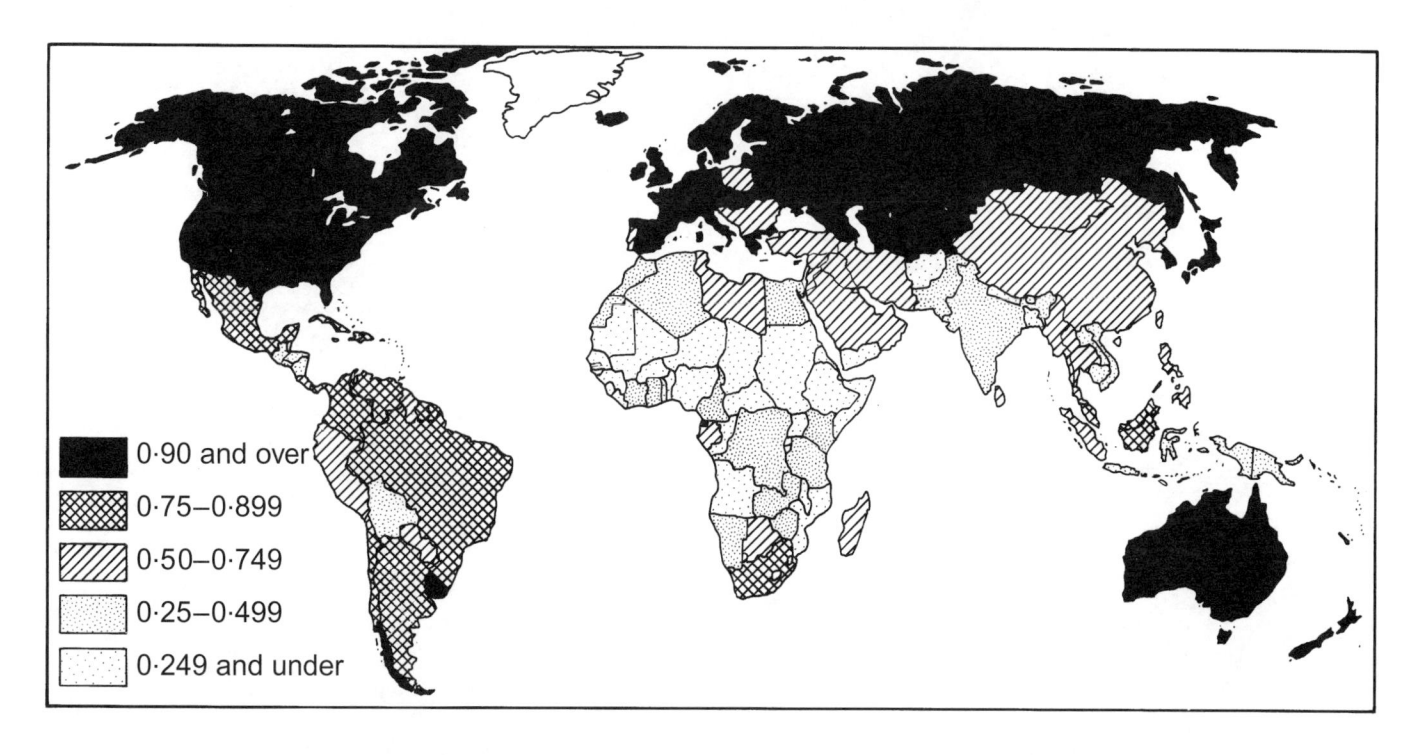

Legend:
- 0·90 and over
- 0·75–0·899
- 0·50–0·749
- 0·25–0·499
- 0·249 and under

The Human Development Index measures development by combining three individual measures. These measures are:

- adult literacy rate;
- life expectancy;
- real Gross Domestic Product (ie what an income will actually buy in a country).

[END OF QUESTION PAPER]

[BLANK PAGE]

HIGHER

2010

OFFICIAL SQA PAST PAPERS | 97 | HIGHER GEOGRAPHY 2010

X208/301

NATIONAL
QUALIFICATIONS
2010

MONDAY, 31 MAY
9.00 AM – 10.30 AM

GEOGRAPHY
HIGHER
Paper 1
Physical and
Human Environments

Six questions should be attempted, namely:

all four questions in **Section A** (Questions 1, 2, 3 and 4);

one question from **Section B** (Question 5 **or** Question 6);

one question from **Section C** (Question 7 **or** Question 8).

Write the numbers of the **six** questions you have attempted in the marks grid on the back cover of your answer booklet.

The value attached to each question is shown in the margin.

Credit will be given for appropriate maps and diagrams, and for reference to named examples.

Questions should be answered in sentences.

Note The reference maps and diagrams in this paper have been printed in black only: no other colours have been used.

SQA

PB X208/301 6/13810

1:50 000 Scale
Landranger Series

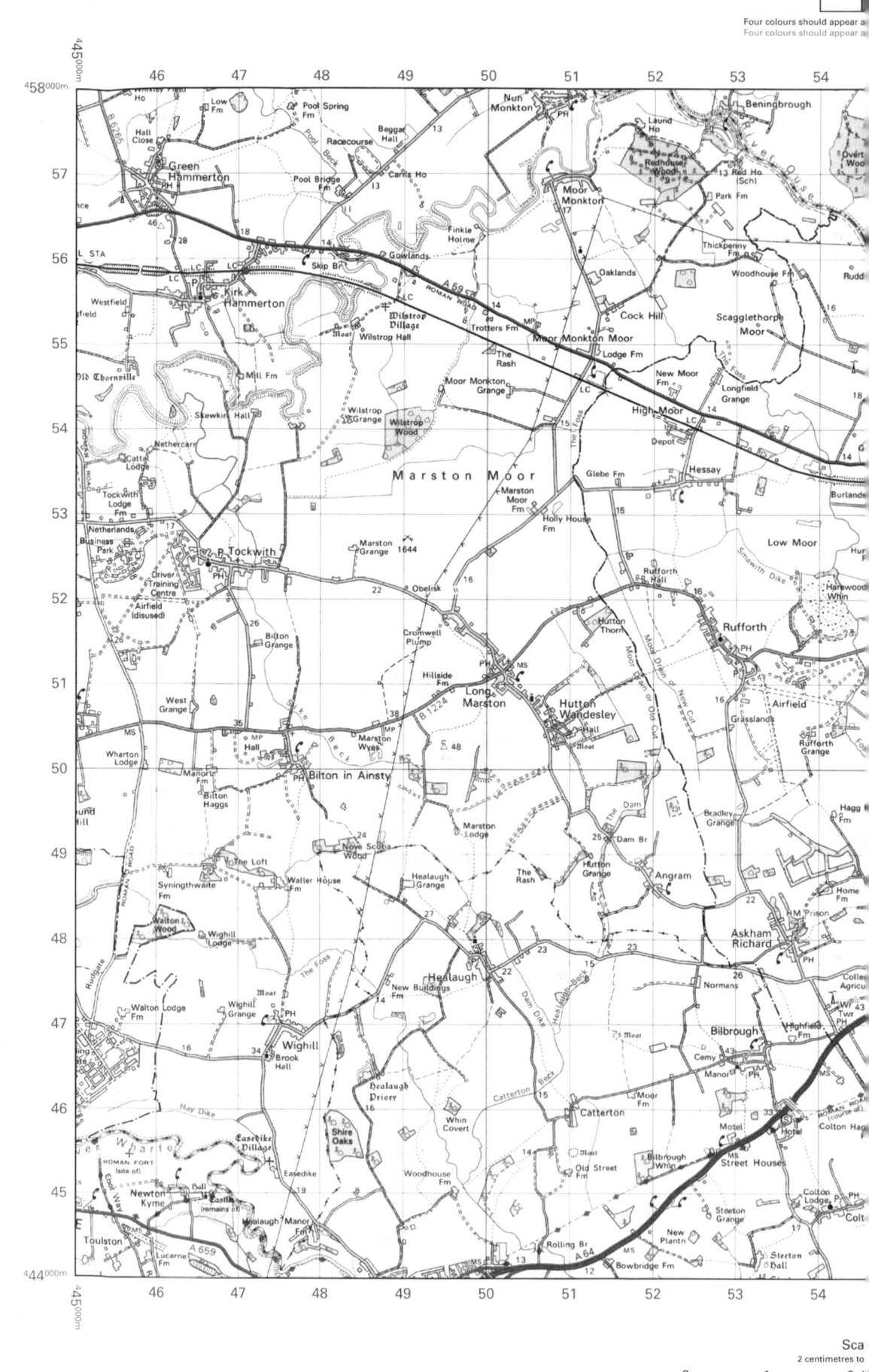

Extract produced by Ordnance Survey 2009. Licence: 100035658
© Crown copyright 2006. All rights reserved.
Ordnance Survey, OS, the OS Symbol and Landranger are registered trademarks of Ordnance Survey, the national mapping agency of Great Britain.
Reproduction in whole or in part by any means is prohibited without the prior written permission of Ordnance Survey. **For educational use only.**

1 kilometre = 0·6214 mile

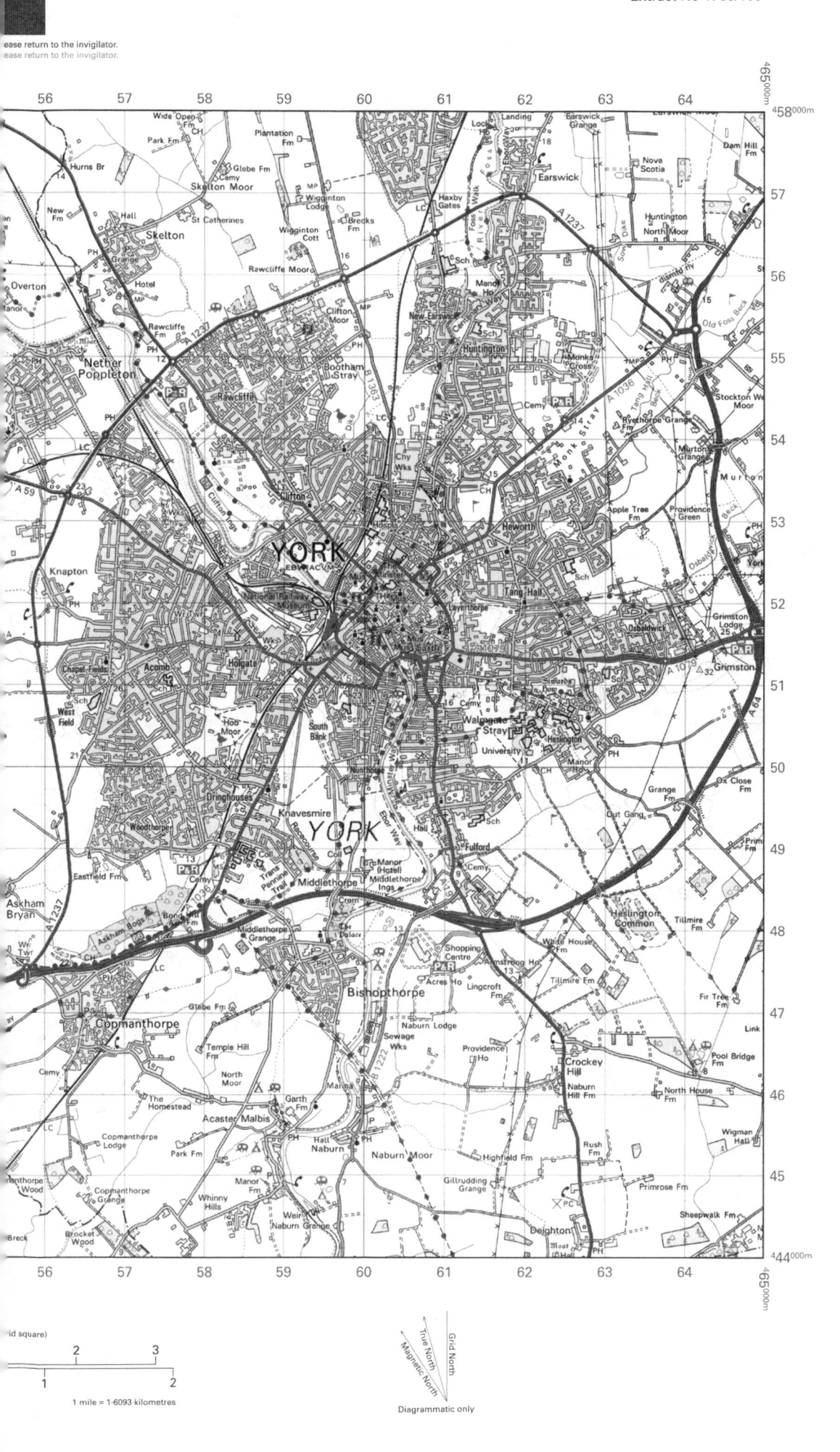

OFFICIAL SQA PAST PAPERS   HIGHER GEOGRAPHY 2010

Extract No 1788/105

SECTION A: Answer ALL questions in this section

Marks

Question 1: Atmosphere

Study Diagram Q1.

(a) **Describe** the **human** factors that may lead to the global temperature projection shown in the diagram. **10**

(b) **Describe** and **explain** the possible consequences of global warming. **10**

Diagram Q1: Global Warming Projection

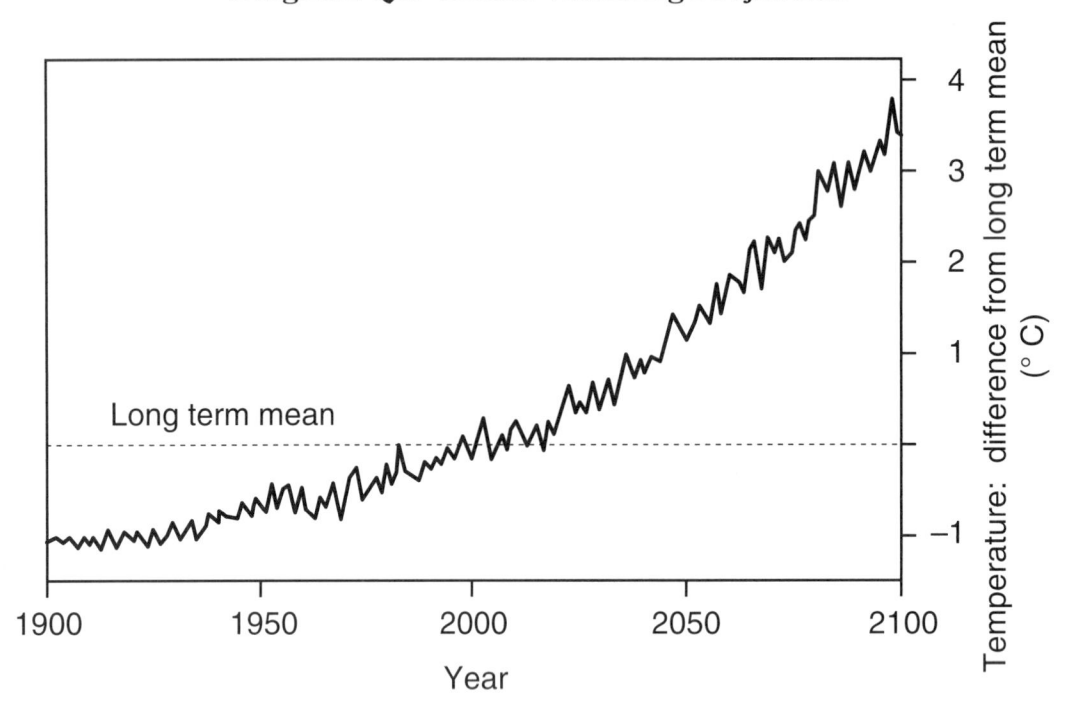

Question 2: Lithosphere

Study Diagram Q2.

Scree is a feature of both glaciated and limestone upland landscapes.

(a) **Describe** and **explain** the conditions **and** processes which encourage the formation of scree slopes. 7

Corries are landscape features in glaciated upland areas.

(b) With the aid of annotated diagrams, **explain** the processes involved in the formation of a corrie. 9

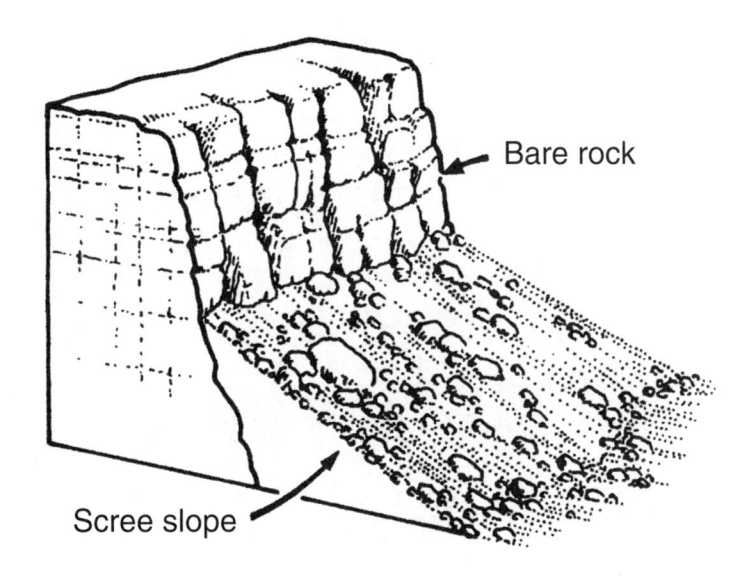

Diagram Q2: Scree

[Turn over

Marks

Question 3: Population Geography

Map Q3 shows the main origins of UK immigrants during 2005/2006.

(a) **Describe** and **suggest reasons** for the patterns shown on Map Q3. **10**

(b) With reference to a migration flow you have studied, **describe** the impact on **either** the donor **or** receiving country. **6**

Map Q3: Main Origins of UK Immigrants 2005/2006

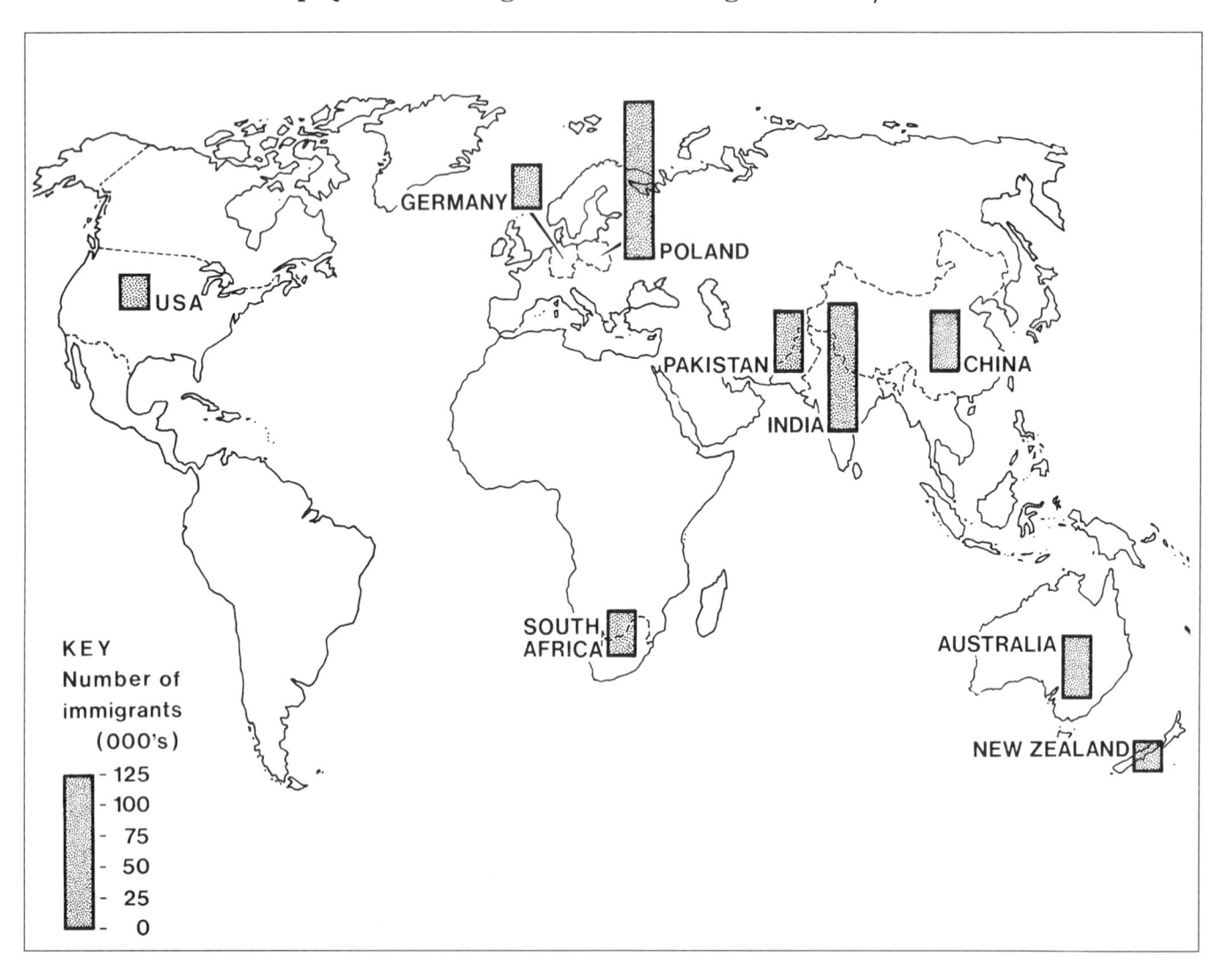

Marks

Question 4: Urban Geography

Study OS Map Extract number 1788/105: York (*separate item*), and Map Q4.

(a) What **map evidence** suggests that the Central Business District of York lies within Area A? **6**

(b) For **either** Area B **or** Area C, **explain** the advantages of its location and environment for its residents. **7**

(c) Using map evidence, **explain** why the southward expansion of York into Area D may create land use conflicts. **7**

Map Q4: Location of urban areas in York

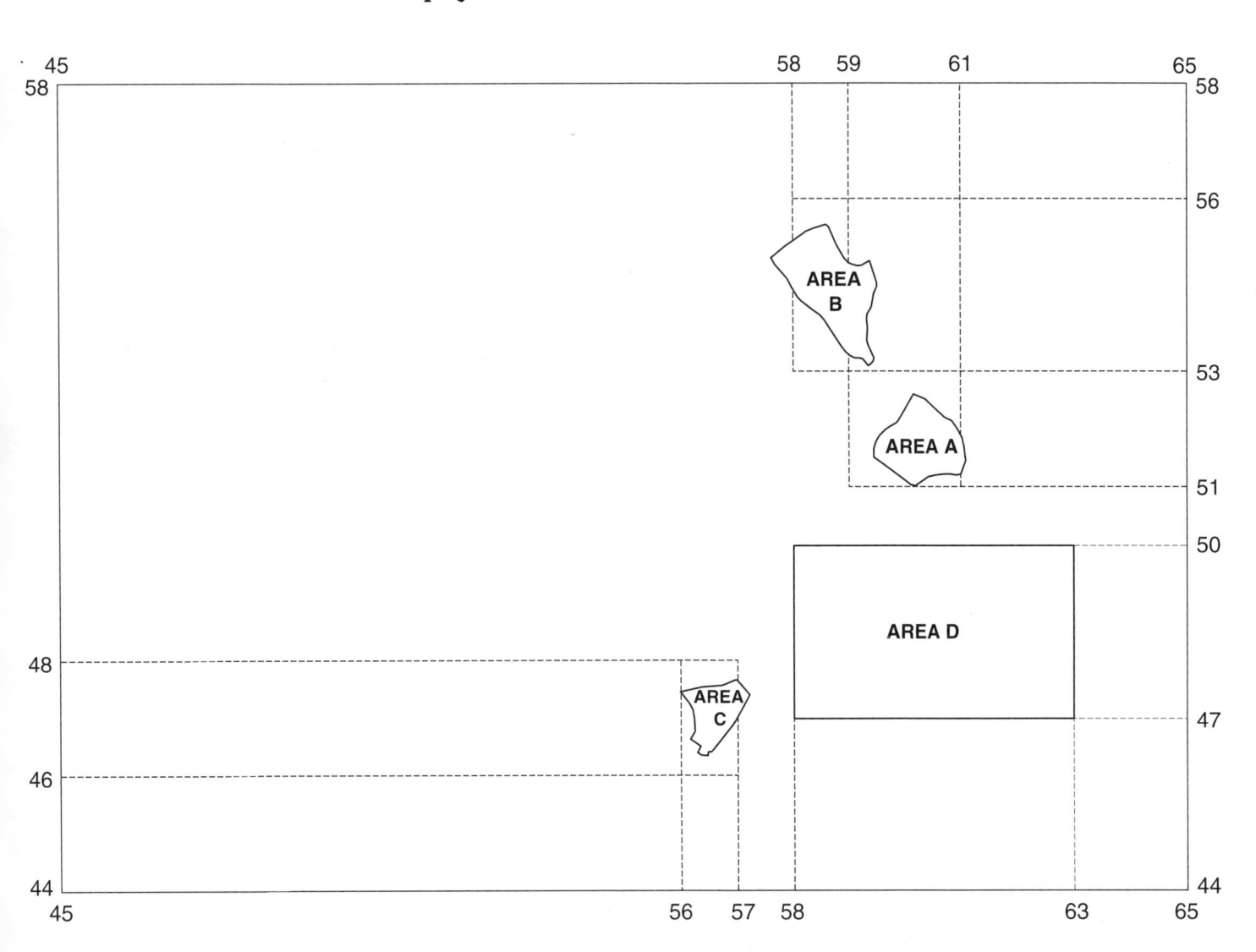

[Turn over

SECTION B: Answer ONE question from this section, ie either Question 5 or Question 6.

Marks

Question 5: Hydrosphere

Study OS Map Extract number 1788/105: York (*separate item*).

(a) Meanders have formed on the River Nidd from GR 450542 to its confluence with the River Ouse GR 513578.

Describe and **explain**, with the aid of a diagram or diagrams, how a meander is formed.   8

(b) Study the OS Map Extract and Map Q5.

"The 2000 floods were the worst in York since records began and the River Ouse reached a height of 5·3 metres above its normal summer level."

(BBC News, November 2000)

With the aid of map evidence, **explain** the physical **and** human factors which may have contributed to the flooding in York after periods of extreme rainfall.   6

Map Q5: Flooded areas of York, 2000

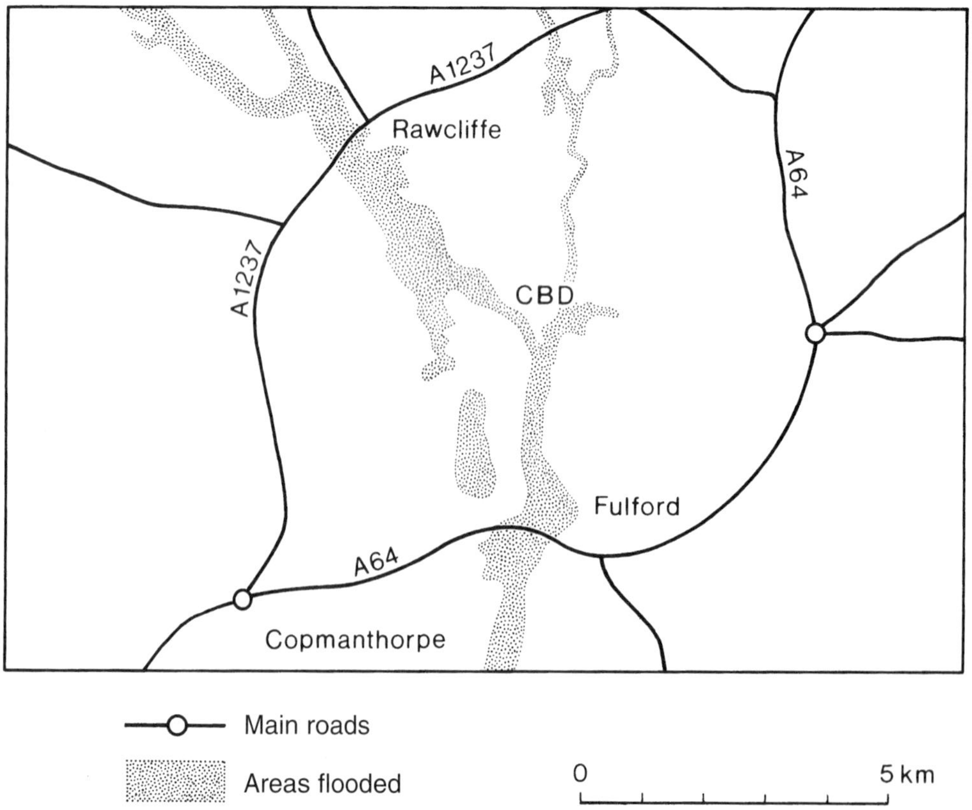

DO NOT ANSWER THIS QUESTION IF YOU HAVE ALREADY ANSWERED QUESTION 5

Question 6: Biosphere

Study Diagram Q6 which shows some of the factors involved in vegetation succession on sand dunes.

Explain why there is a change in vegetation cover and species as you move inland from the beach. You should refer to named plant species in your answer.

14

Diagram Q6: Factors involved in sand dune succession

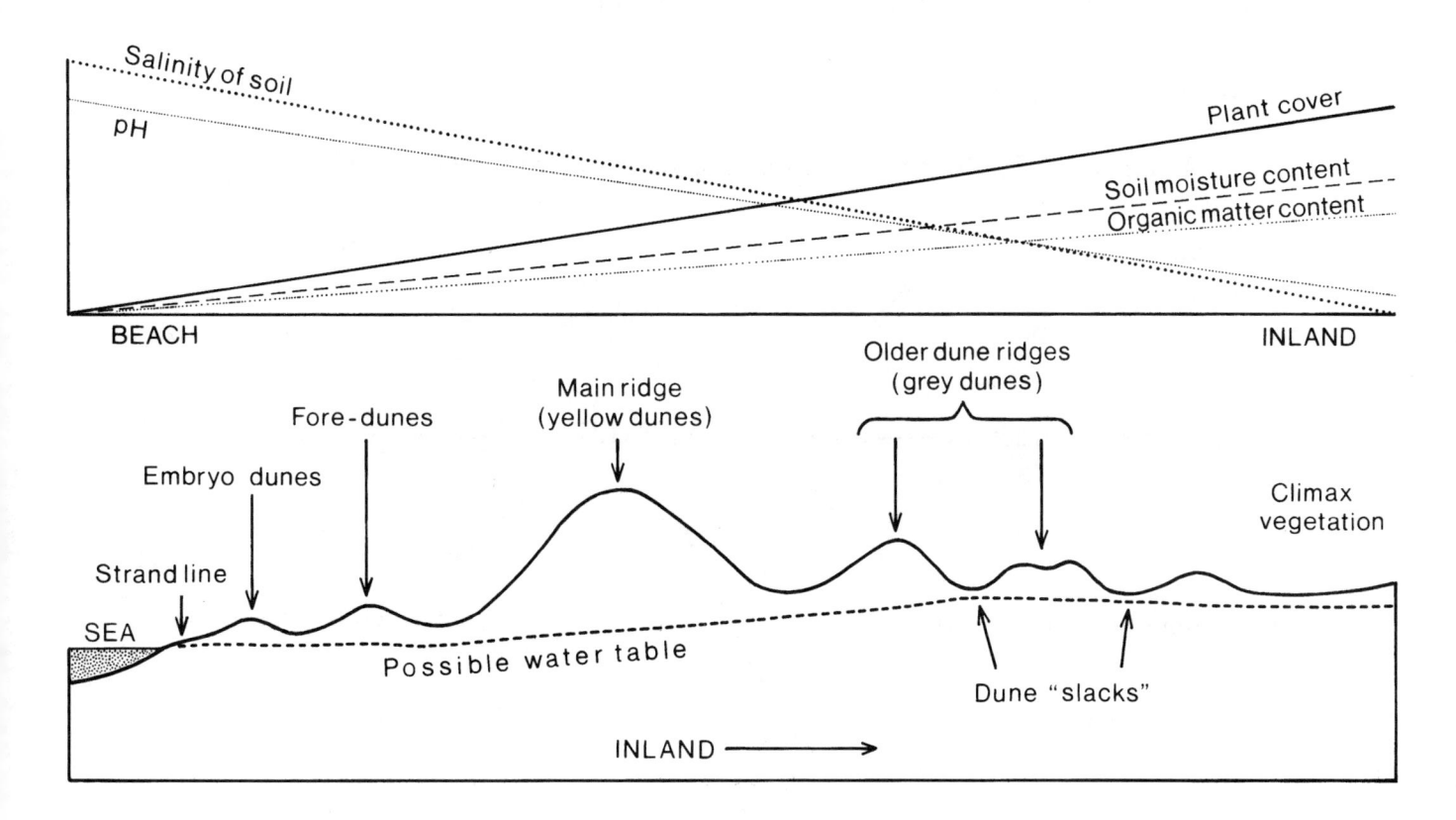

[Turn over

SECTION C: Answer ONE question from this section, ie either Question 7 or Question 8.

Question 7: Rural Geography

"Mechanisation has led to major changes in commercial arable farming."

(a) (i) **Suggest why** farmers have invested in increased mechanisation.

   (ii) **Explain** the impact of increased mechanisation on the environment.   **8**

(b) Study Diagram Q7 which shows some of the other recent changes in commercial arable farming.

   Describe and **explain two** of the changes shown in Diagram Q7.   **6**

Diagram Q7: Recent changes in commercial arable farming

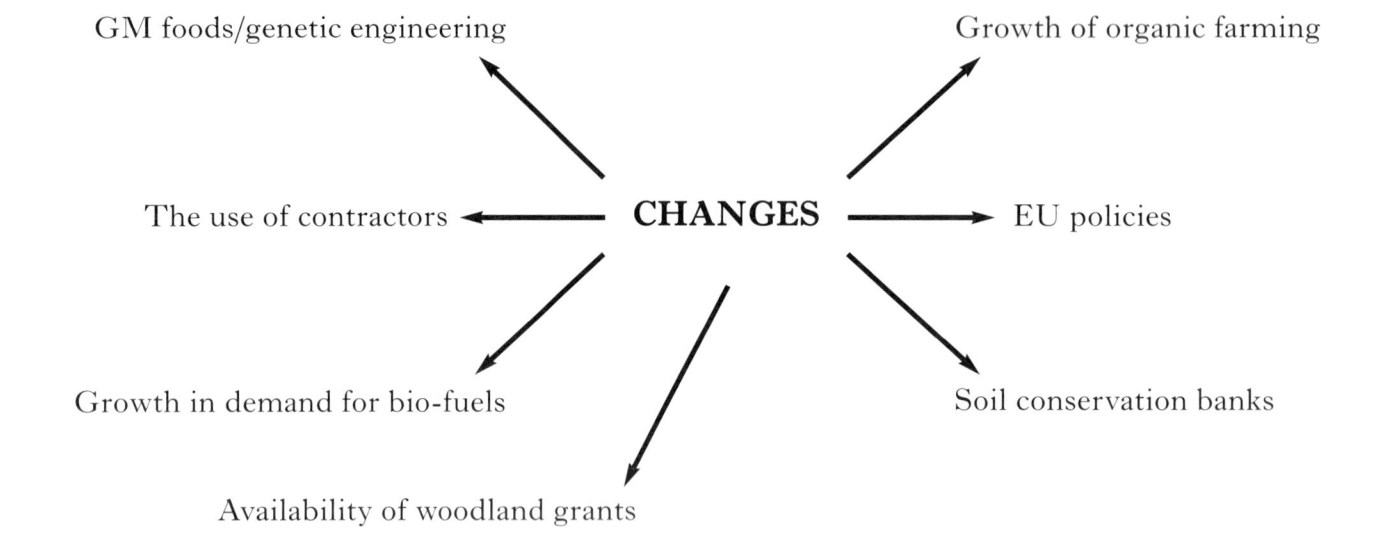

- GM foods/genetic engineering
- Growth of organic farming
- The use of contractors → CHANGES → EU policies
- Growth in demand for bio-fuels
- Soil conservation banks
- Availability of woodland grants

DO NOT ANSWER THIS QUESTION IF YOU HAVE ALREADY ANSWERED QUESTION 7

Question 8: Industrial Geography

With reference to named examples within an area of industrial decline in the European Union you have studied:

(i) **give reasons** for the industrial decline; and 8

(ii) **describe** the socio-economic impacts of the closure of such industries on the local population and the surrounding area. 6

[END OF QUESTION PAPER]

[BLANK PAGE]

X208/303

NATIONAL
QUALIFICATIONS
2010

MONDAY, 31 MAY
10.50 AM – 12.05 PM

GEOGRAPHY
HIGHER
Paper 2
Environmental
Interactions

Answer any **two** questions.

Write the numbers of the **two** questions you have attempted in the marks grid on the back cover of your answer booklet.

The value attached to each question is shown in the margin.

Credit will be given for appropriate maps and diagrams, and for reference to named examples.

Questions should be answered in sentences.

Note   The reference maps and diagrams in this paper have been printed in black only: no other colours have been used.

Marks

Question 1 (Rural Land Resources)

(a) With the aid of annotated diagrams, **describe** and **explain** the physical features associated with the formation of coastal landscapes. You should refer to both erosion **and** deposition features in your answer. **20**

(b) For any named coastal area you have studied, **describe** how this landscape has provided a variety of socio-economic opportunities. **10**

(c) Study Diagram Q1A and Map Q1B.

One example of a land use conflict is the proposed leisure/housing development at the Menie Estate in Aberdeenshire. Part of this development takes place on a protected sand dune area designated as an SSSI. (SSSI = Site of Special Scientific Interest.)

Discuss the advantages **and** disadvantages of developments such as this on the local people and the environment. **10**

(d) For any named coastal **or** upland area you have studied, **describe** the measures taken to resolve environmental conflicts and **comment on** their effectiveness. **10**

(50)

Diagram Q1A: News Reports on the Proposals for the Menie Estate

"Business leaders have joined forces to urge the Scottish Government to give the go-ahead to US billionaire Donald Trump's plans for a golf resort . . . Mr Trump hopes to build a resort featuring two championship golf courses, a five-star hotel, 950 holiday homes and 500 private houses at the Menie Estate in Aberdeenshire."

(*The Herald* 19/6/08)

"The value of Menie Links as part of the Foveran Links SSSI cannot be understated. It is the most dynamic, most rapidly moving and largest area of bare sand in this area of Scotland. It is quite simply the jewel in the crown of the SSSI areas of bare sand in this area of Scotland and therefore the jewel in the crown of the UK resource."

(Scottish Natural Heritage (SNH) expert—*The Herald* 19/6/08)

RSPB Scotland objected to the Trump International application because . . . the developer's own Environmental Statement acknowledges that there will be very significant adverse effects on habitats and biodiversity—the mobile dunes, which form one of the main qualifying features of the Foveran Links SSSI, will be destroyed.

(http://www.rspb.org.uk/ourwork/conservation/sites/scotland/menie.asp)

(RSPB = Royal Society for the Protection of Birds)

Question 1 – continued

Map Q1B: Proposed developments at the Menie Estate

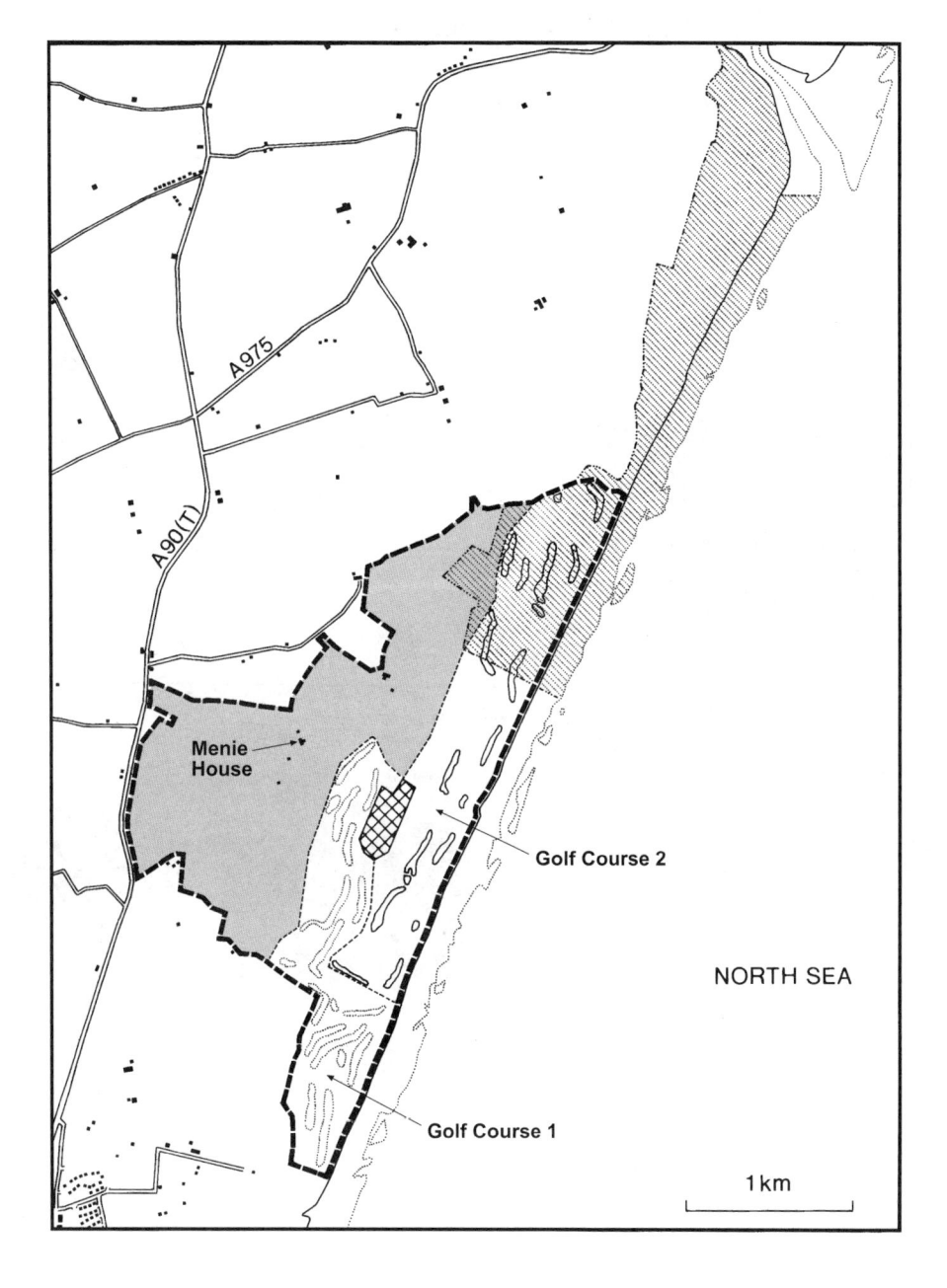

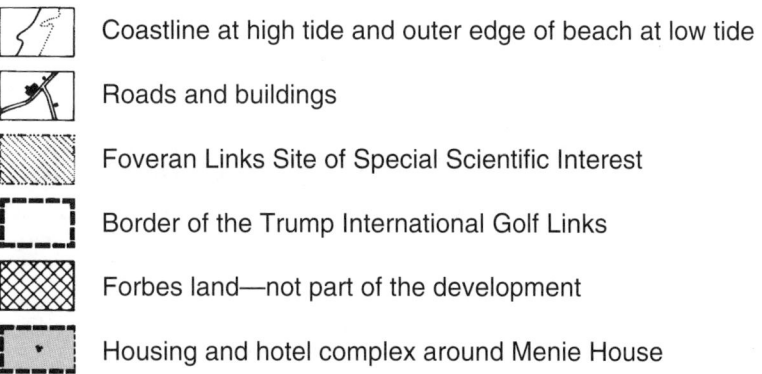

A975
A90(T)
Menie House
Golf Course 2
Golf Course 1
NORTH SEA
1 km
Coastline at high tide and outer edge of beach at low tide
Roads and buildings
Foveran Links Site of Special Scientific Interest
Border of the Trump International Golf Links
Forbes land—not part of the development
Housing and hotel complex around Menie House

Question 2 (Rural Land Degradation)

(a) Study Diagram Q2A.

Describe and **explain** the processes of soil erosion by water. 8

Diagram Q2A: Erosion by water

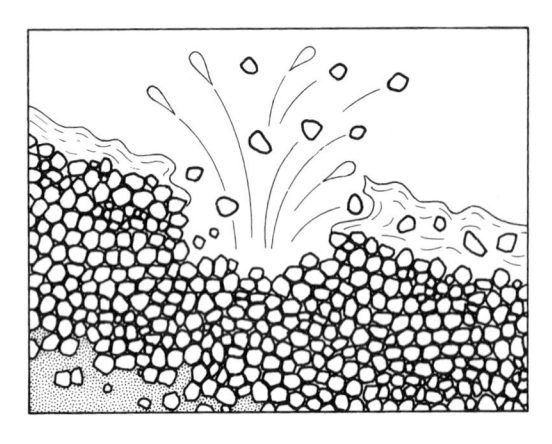

Rainsplash

Sheet Erosion

Rill Erosion

Gully Erosion

Marks

Question 2 – continued

(b) Study Table Q2B.

Describe and **explain** how human activities have caused land degradation in North America **and either** Africa north of the Equator **or** the Amazon Basin. 16

Table Q2B: Percentage of the agricultural land which has been degraded

Region	% degraded
North America	26
Africa	65
South America	45

(c) Referring to named locations in **either** Africa north of the Equator **or** the Amazon Basin, **describe** the impact of land degradation on the people and economy. 10

(d) Referring to named locations in North America you have studied:

(i) **describe** and **explain** the ways in which farmers have adjusted their farming methods to reduce the risk of soil erosion; and

(ii) **comment** on the effectiveness of these methods. 16

(50)

[Turn over

Marks

Question 3 (River Basin Management)

(a) *"The Myitsore hydro-electric project was started in 2008 to manage the flow of the Irrawaddy River in northern Myanmar (Burma)."*

Study Map Q3A and Diagrams Q3A, Q3B and Q3C.

  (i) **Describe** and **account for** the pattern of river flow before the Myitsore Project started.

  (ii) **Describe** and **explain** the need for water management in the Irrawaddy River in Myanmar. **16**

(b) For the Myitsore Dam **or** any named dam you have studied in Africa **or** North America **or** Asia, **describe** and **explain** the physical factors which should be considered when selecting the site for the dam and its associated reservoir. **10**

(c) **Describe** and **account for** the social, economic and environmental benefits **and** adverse consequences of a named water management project in Africa **or** North America **or** Asia. **24**

(50)

Diagram Q3A: Monthly discharge of the Irrawaddy River at Myitsore before the HEP scheme

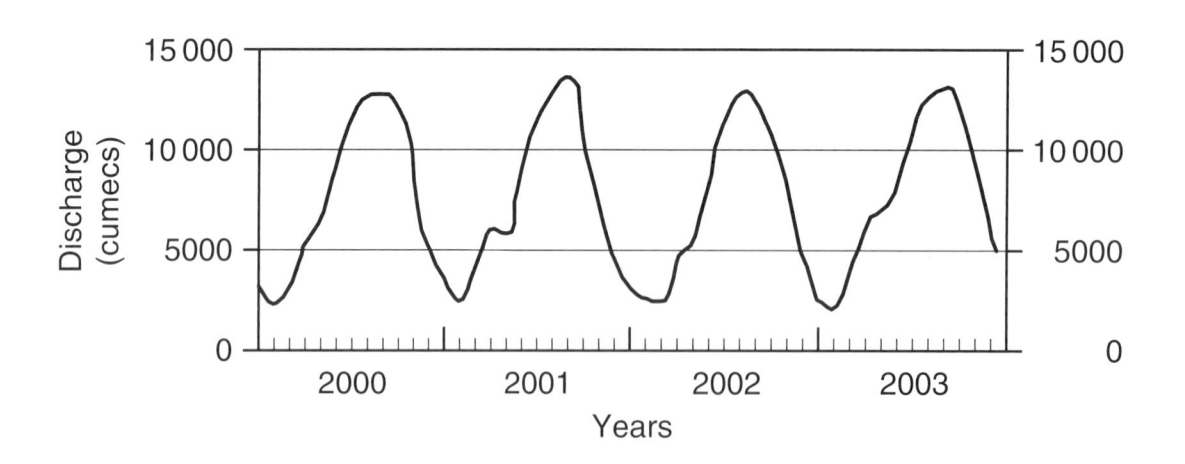

Question 3 – continued

Map Q3A: Irrawaddy River in Myanmar

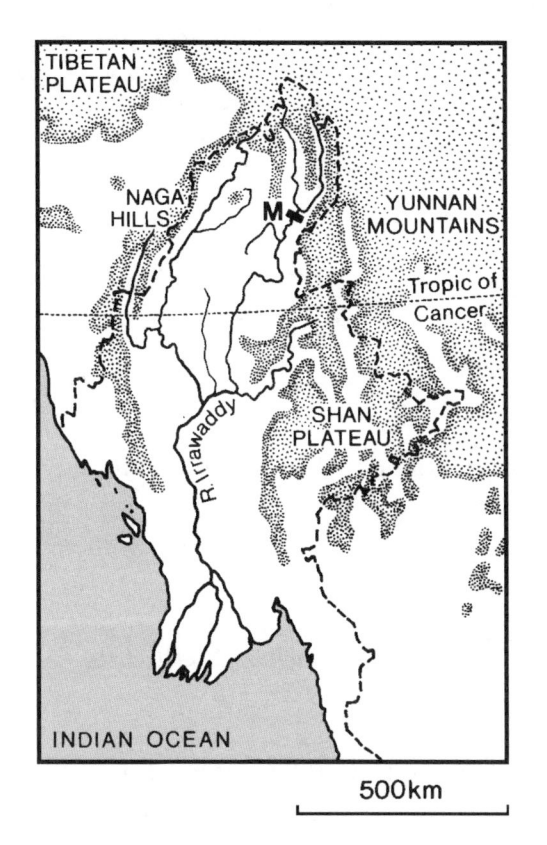

M↝ Myitsore Dam

Land over 1000 metres

Diagram Q3B: Myitsore—Climate Graph

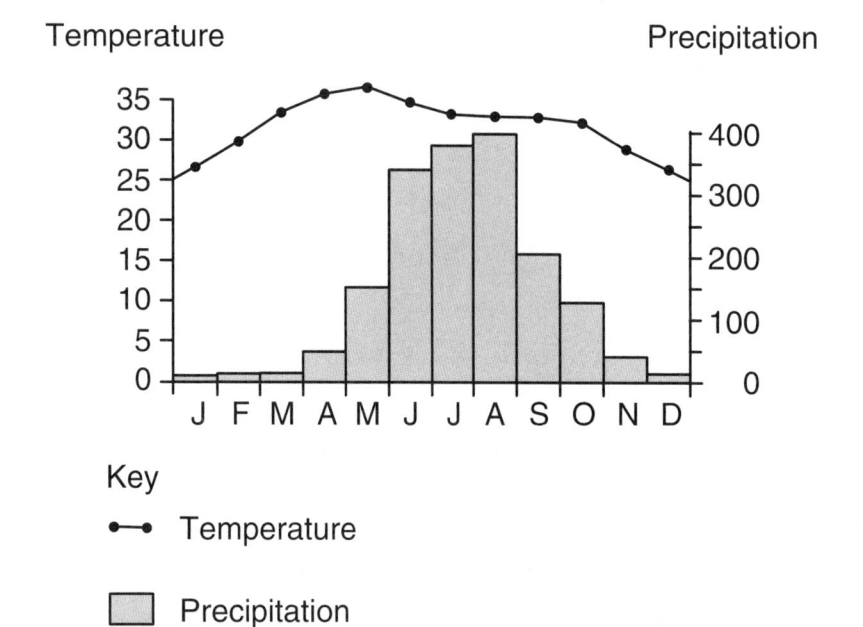

Key
- —●— Temperature
- ▨ Precipitation

Diagram Q3C: Projected population change in Myanmar

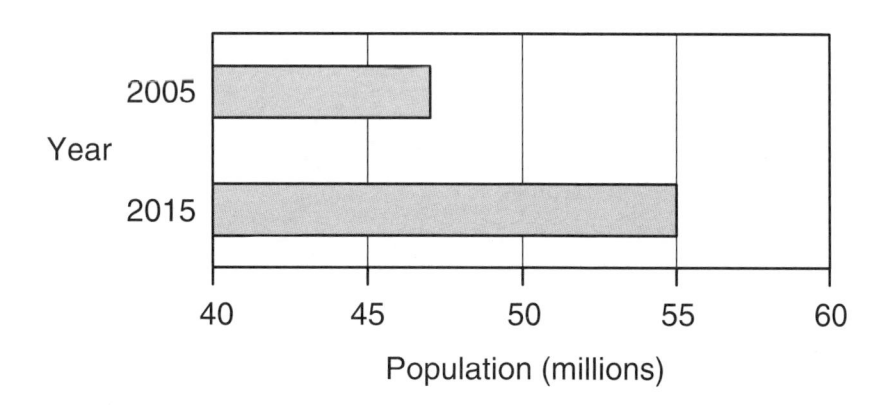

[Turn over

Marks

Question 4 (Urban Change and its Management)

(a) *"A megacity is defined as a city with over 10 million people."*

Study Diagram Q4A (on *Page nine*).

Describe the changes in the number and world distribution of megacities from 1975 to 2015.　　**8**

(b) Study Diagrams Q4A and Q4B (on *Pages nine* and *ten*).

For Mexico City **or** any other named city which you have studied in a Developing Country:

(i) **explain** the growth of your chosen city in terms of rural push/urban pull factors;　　**10**

(ii) **describe** the socio-economic and environmental problems which have resulted from this rapid growth.　　**12**

(c) Study Map Q4 (on *Page ten*).

"Urban sprawl has been seen as a problem since the 1930s and regions such as South-East England have come under increasing pressure."

Referring to London **or** any other named city you have studied in a Developed Country:

(i) **explain** the reasons for urban sprawl;

(ii) **outline** the problems caused by this growth;

(iii) select **one** problem identified in part (ii) above and **explain** the ways in which the city has tried to resolve this problem.　　**20**

(50)

[X208/303]　　*Page eight*

Question 4 – continued

Diagram Q4A: The Growth of Megacities 1975–2015

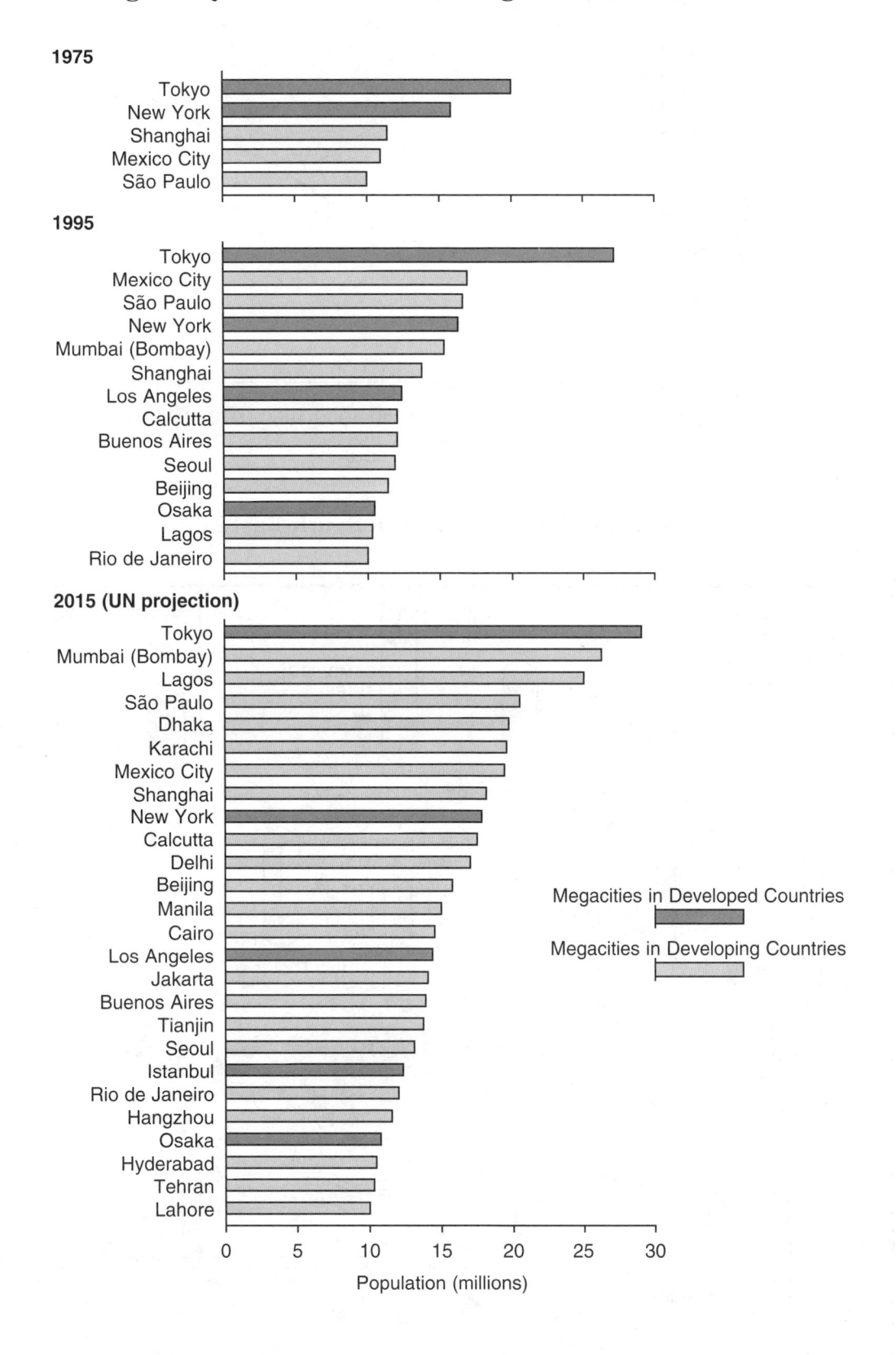

1975

City	Population (millions)
Tokyo	~20
New York	~15.5
Shanghai	~11
Mexico City	~10.5
São Paulo	~10

1995

City	Population (millions)
Tokyo	~27
Mexico City	~16.5
São Paulo	~16
New York	~16
Mumbai (Bombay)	~15
Shanghai	~13.5
Los Angeles	~12
Calcutta	~12
Buenos Aires	~12
Seoul	~11.5
Beijing	~11
Osaka	~10.5
Lagos	~10
Rio de Janeiro	~10

2015 (UN projection)

City	Population (millions)
Tokyo	~29
Mumbai (Bombay)	~26
Lagos	~25
São Paulo	~20.5
Dhaka	~19.5
Karachi	~19.5
Mexico City	~19
Shanghai	~18
New York	~17.5
Calcutta	~17.5
Delhi	~17
Beijing	~15.5
Manila	~15
Cairo	~14.5
Los Angeles	~14.5
Jakarta	~14
Buenos Aires	~14
Tianjin	~13.5
Seoul	~13
Istanbul	~12
Rio de Janeiro	~12
Hangzhou	~11.5
Osaka	~11
Hyderabad	~10.5
Tehran	~10.5
Lahore	~10

Legend: Megacities in Developed Countries; Megacities in Developing Countries

[Turn over

Question 4 – continued

Diagram Q4B: Changes in balance of urban/rural population in Mexico

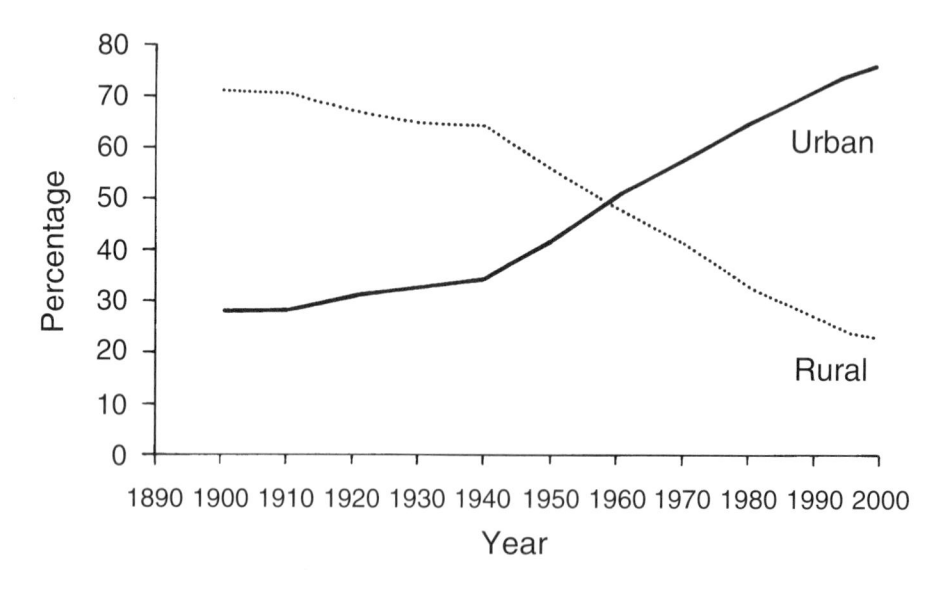

Map Q4: Urban Sprawl in South-East England

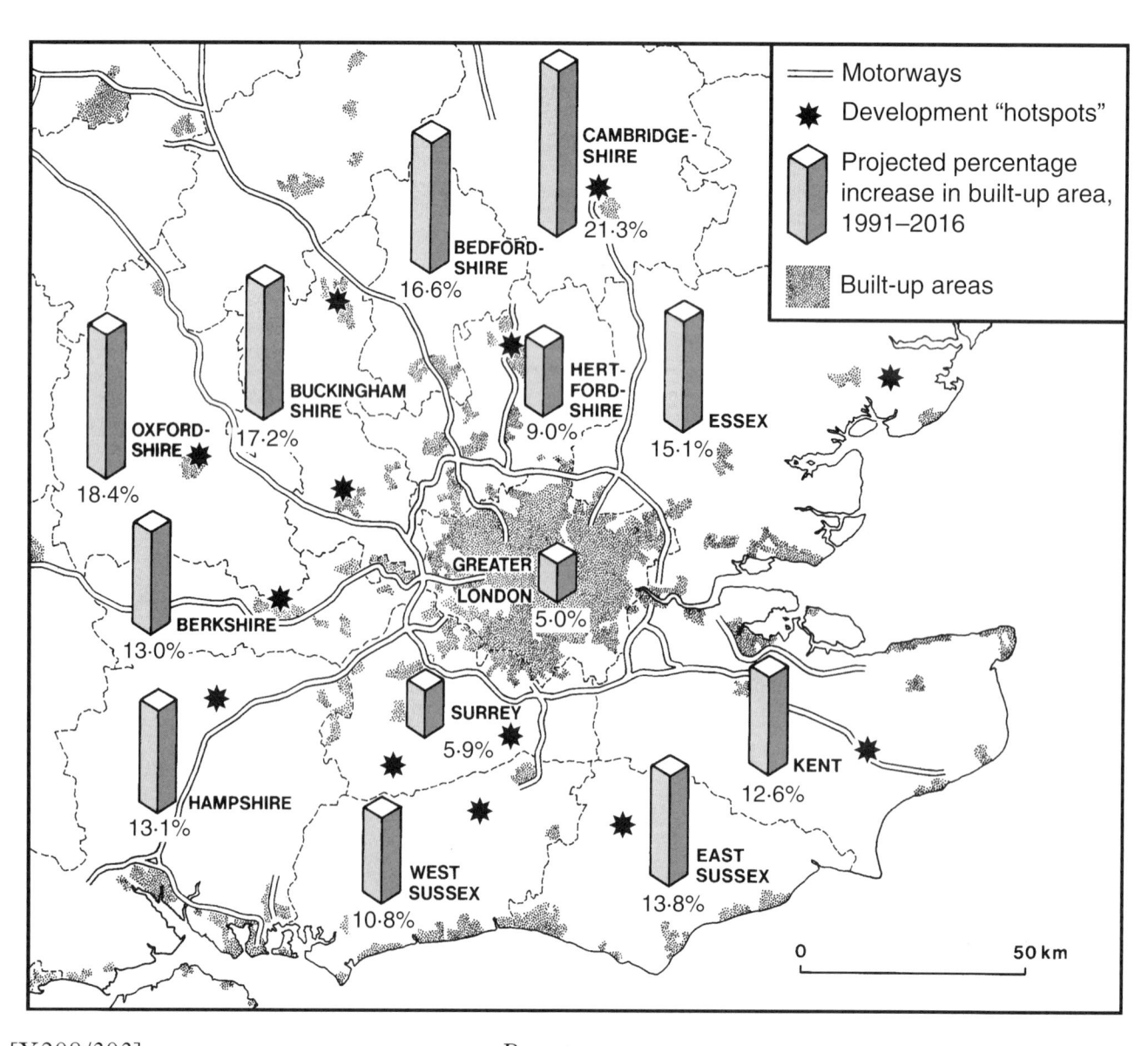

[Turn over for Question 5 on *Page twelve*

Marks

Question 5 (European Regional Inequalities)

(a) Study Map Q5A.

"*Convergence Regions are areas designated as requiring most financial assistance across the European Union (EU).*"

(i) **Describe** the distribution of the Convergence Regions. — 8

(ii) In 2008, the EU budget for promoting growth across the least developed regions was 47 billion Euros. **Discuss** ways in which less prosperous regions can receive help from the EU. — 10

(b) "*The **North-South divide** refers to the economic and cultural differences between southern England and the rest of the United Kingdom.*"

Study Map Q5B and Table Q5.

(i) To what extent does the data provide evidence of regional inequalities within the UK? — 10

(ii) **Describe** and **explain** the physical and human factors that have led to the regional inequalities within the UK. — 14

(iii) **Describe** the steps taken by the UK government agencies to reduce regional inequalities. — 8

(50)

Map Q5A: EU Convergence Regions

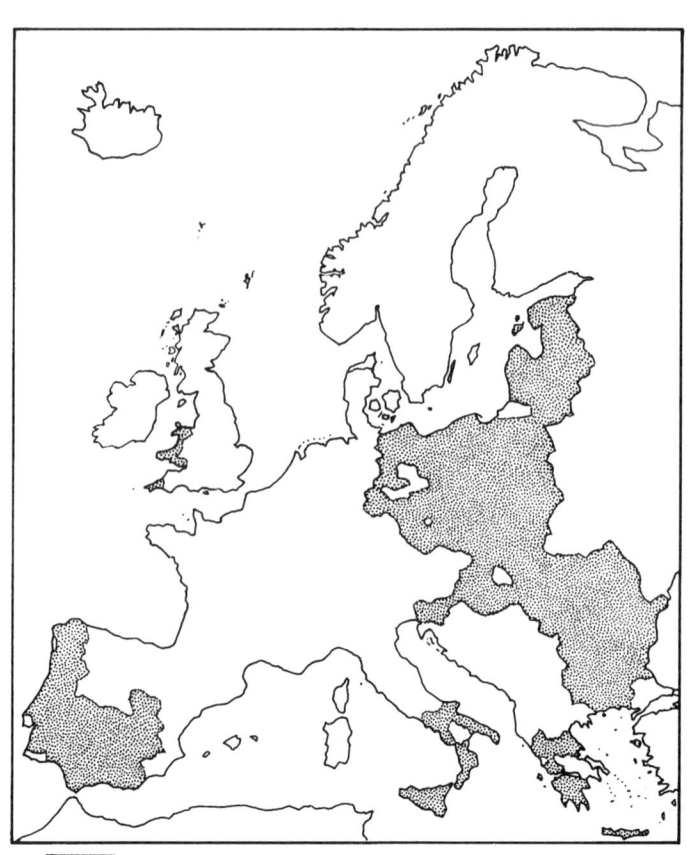

Convergence Regions receiving most financial aid

Question 5 – continued

Map Q5B: UK statistical regions

Table Q5: UK average values

	Population change 1996–2006 %	Average house prices (Nov 2008) £1000	Gross disposable household income (2006) (UK average =100)	Working age population with no qualifications (%) (2006)
UK average	4·3	203	100	13
Scotland	0·0	160	95	13
Northern Ireland	5·1	226	87	22
Wales	2·4	158	89	17
NW England	0·0	158	92	15
NE England	−1·1	148	86	14
West Midlands	2·3	176	91	17
Yorks & Humber	2·7	157	93	15
East Midlands	5·5	166	95	13
East England	6·6	204	107	12
SE England	6·7	271	113	9
London	7·4	382	120	12
SW England	6·9	229	100	9

[X208/303] *Page thirteen* [Turn over

Question 6 (Development and Health)

(a) **Suggest reasons** for the wide variations in development which exist **between** Developing Countries. You should refer to named countries you have studied. **12**

(b) Study Table Q6A, and Maps Q6A, Q6B, Q6C and Q6D.

"*Life expectancy in Chad is only 47 years.*"

Suggest the physical **and** human factors which may have led to this low life expectancy. **12**

(c) Study Map Q6C.

Chad and many other developing countries have been affected by water-related diseases including malaria, cholera and bilharzia/schistosomiasis.

Select **one** of the above diseases.

  (i) **Describe** the physical **and** human factors which put people at risk of contracting the disease. **8**

  (ii) **Describe** and **explain** the measures that can be taken to combat the disease. **14**

  (iii) **Explain** the benefits to a Developing Country of controlling the disease. **4**

(50)

Table Q6A: Selected development indicators for Chad

Indicator	
GDP per capita ($US)	1500
Birth rate per 1000	42
Infant mortality per 1000 live births	100
% land surface for arable farming	3
Adult literacy rate (%)	25
% population with HIV/AIDS	4·8

Question 6 – continued

Map Q6A: Map of Chad

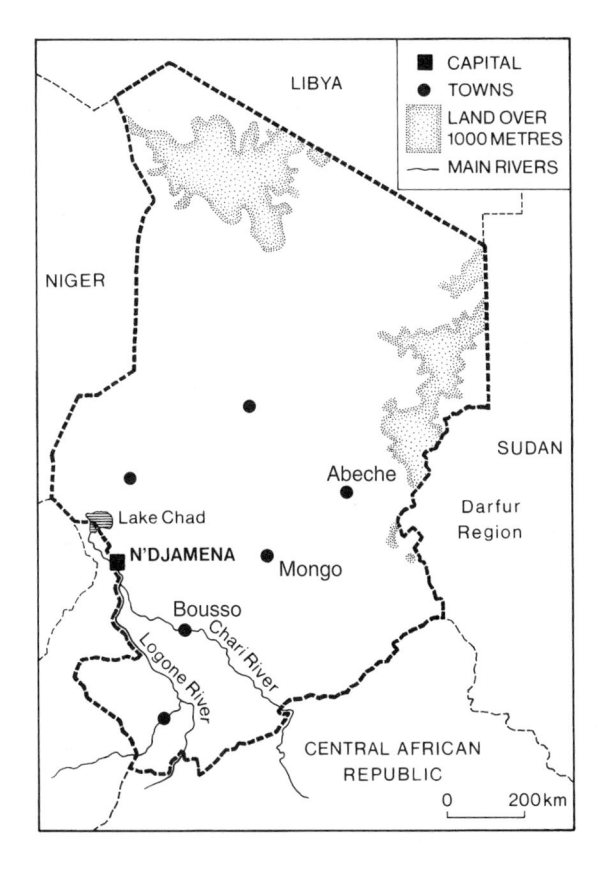

Map Q6B: Bio-climatic zones of Chad

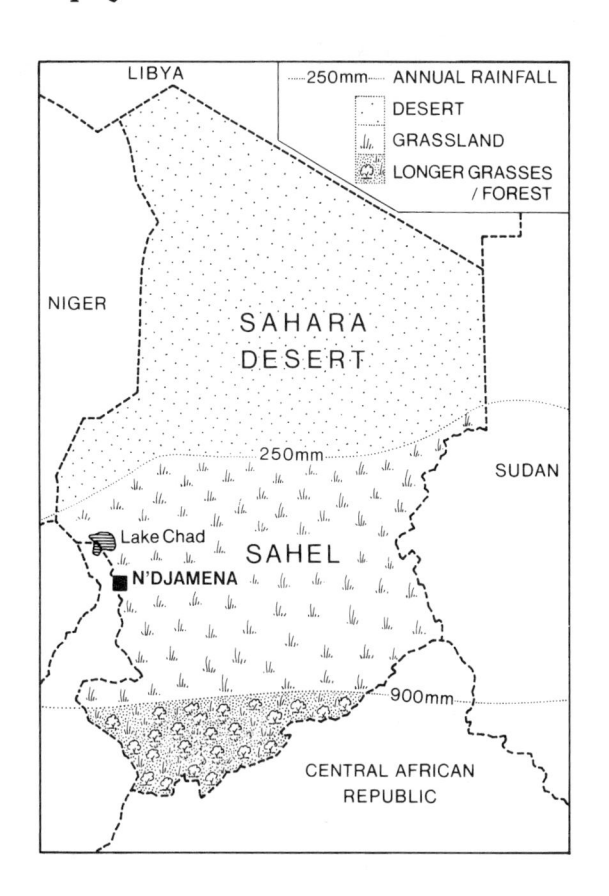

Map Q6C: Disease in Chad

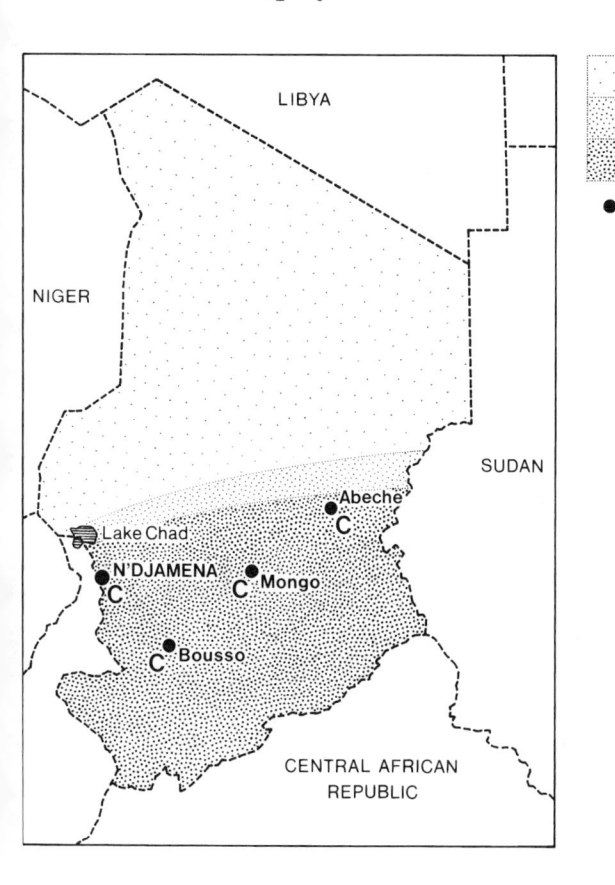

Map Q6D: Location of Chad within Africa

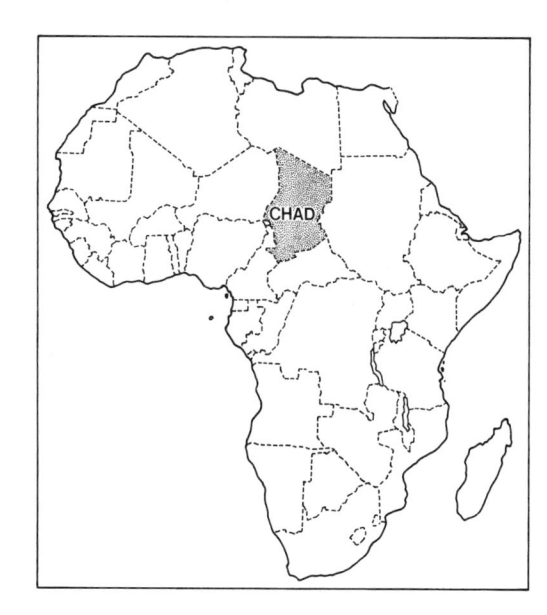

[END OF QUESTION PAPER]

[BLANK PAGE]

HIGHER

2011

[BLANK PAGE]

X208/301

| NATIONAL QUALIFICATIONS 2011 | TUESDAY, 24 MAY 9.00 AM – 10.30 AM | GEOGRAPHY HIGHER Paper 1 Physical and Human Environments |

Six questions should be attempted, namely:

all four questions in **Section A** (Questions 1, 2, 3 and 4);

one question from **Section B** (Question 5 **or** Question 6);

one question from **Section C** (Question 7 **or** Question 8).

Write the numbers of the **six** questions you have attempted in the marks grid on the back cover of your answer booklet.

The value attached to each question is shown in the margin.

Credit will be given for appropriate maps and diagrams, and for reference to named examples.

Questions should be answered in sentences.

Note   The reference maps and diagrams in this paper have been printed in black only: no other colours have been used.

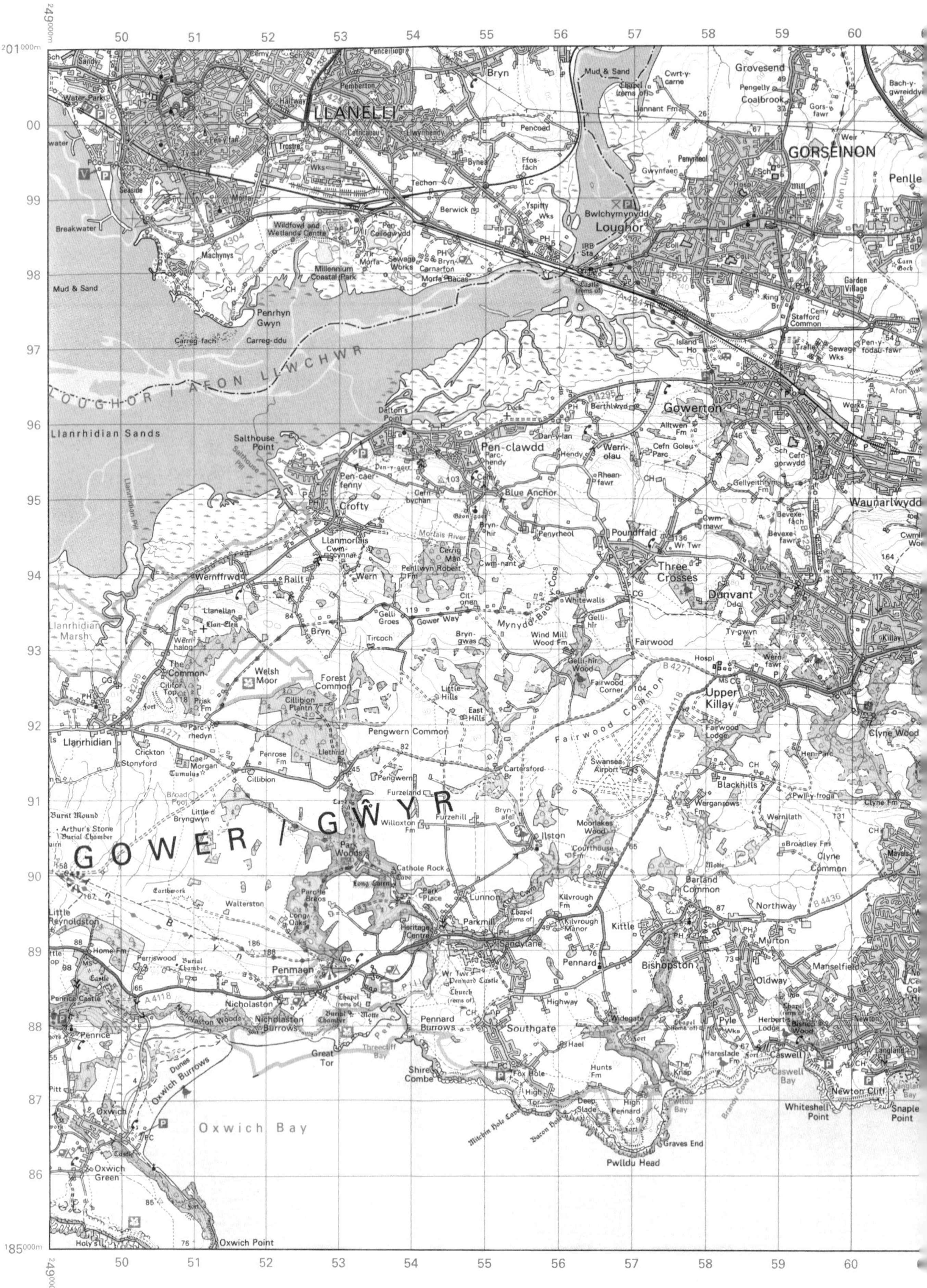

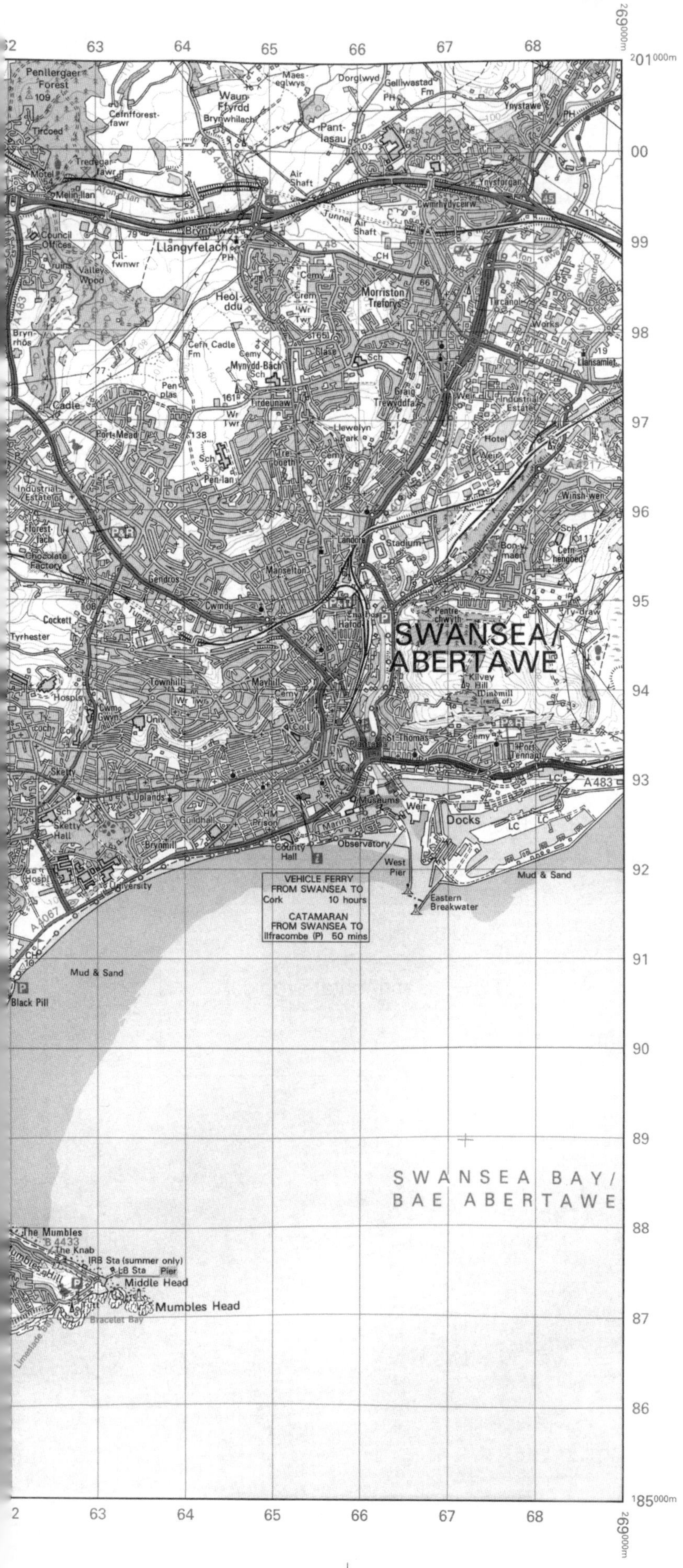

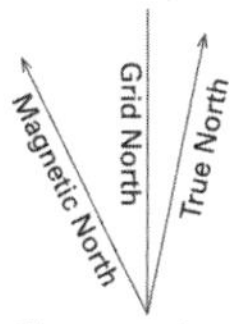

Diagrammatic only

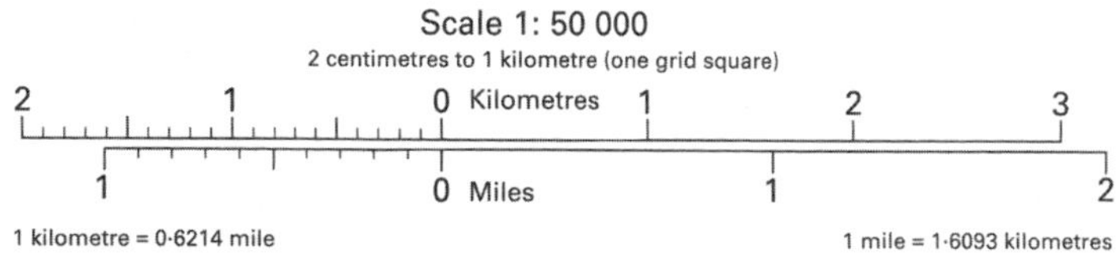

Question 1: Atmosphere

Study Maps Q1A.

(a) **Describe** the origin, nature and characteristics of the Maritime Tropical and Continental Tropical air masses. 6

Maps Q1A: Location of selected air masses and the ITCZ in January and July

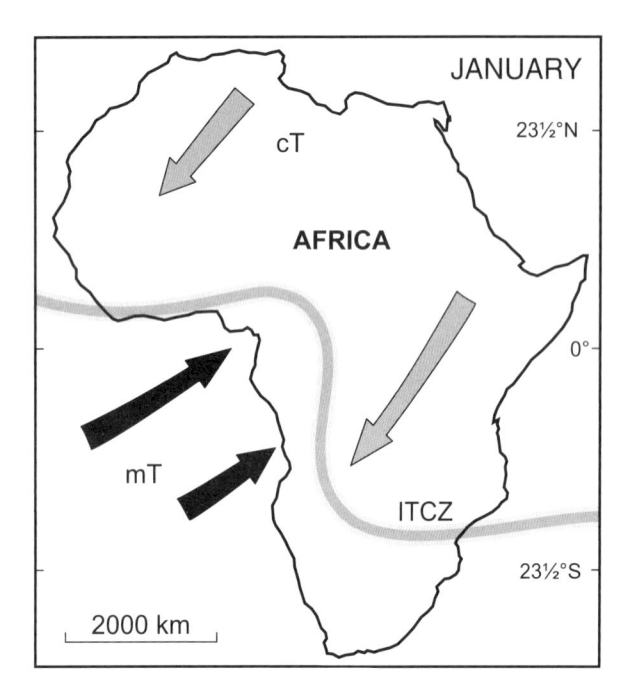

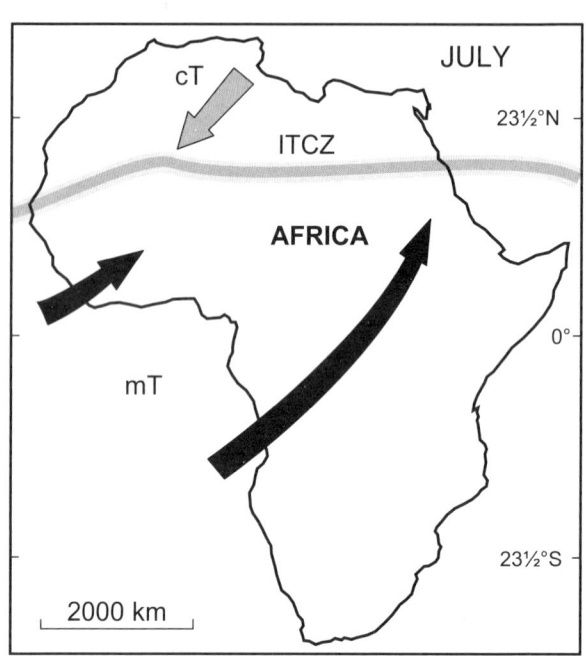

Key:

mT = Maritime Tropical   cT = Continental Tropical

ITCZ = Inter Tropical Convergence Zone

Question 1: Atmosphere (continued)

Study Maps Q1A and Q1B and Diagram Q1.

(b) **Describe** and **explain** the variation in rainfall within West Africa. **12**

Map Q1B: West Africa

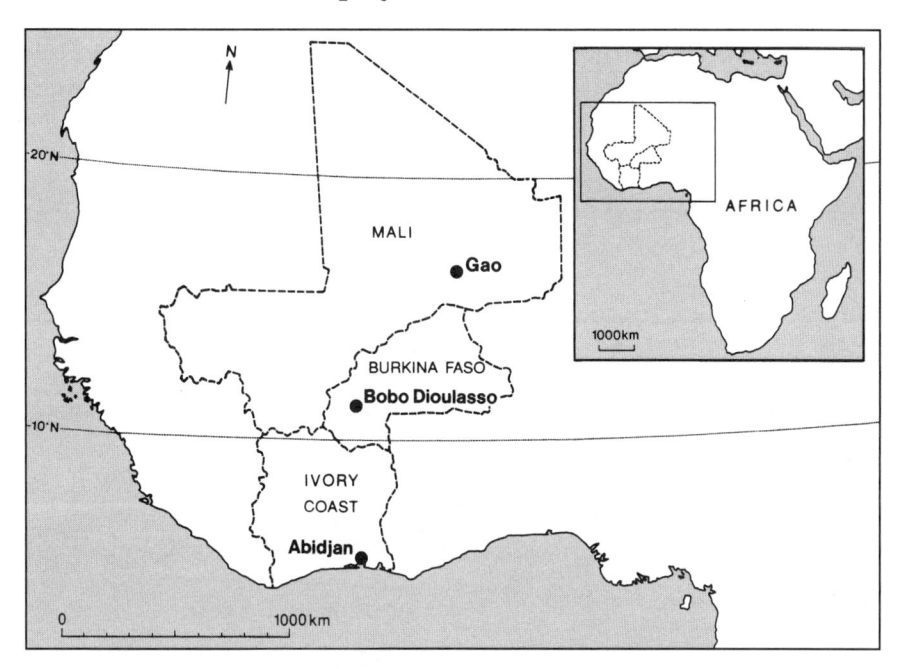

Diagram Q1: Average Monthly Rainfall/Days with Precipitation

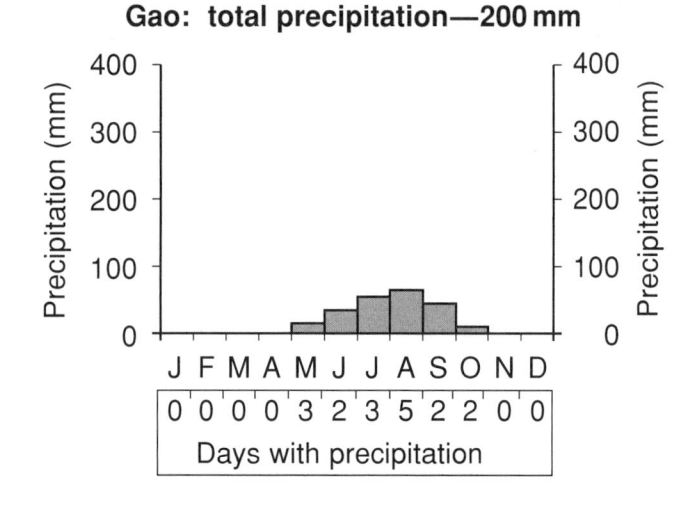

Gao: total precipitation—200 mm

Bobo-Dioulasso: total precipitation—1000 mm

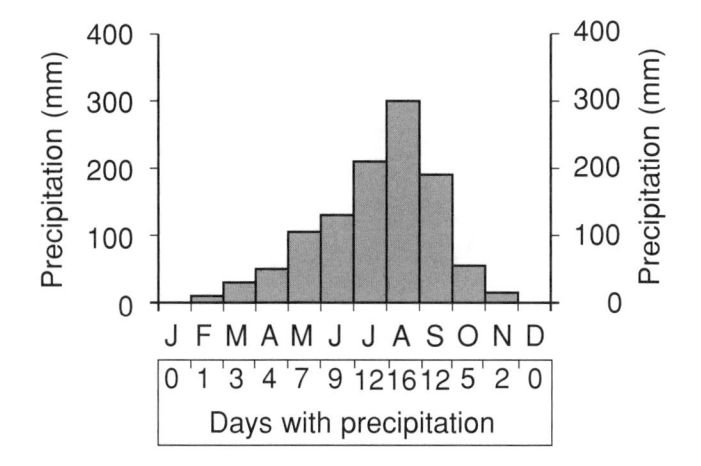

Abidjan: total precipitation—1700 mm

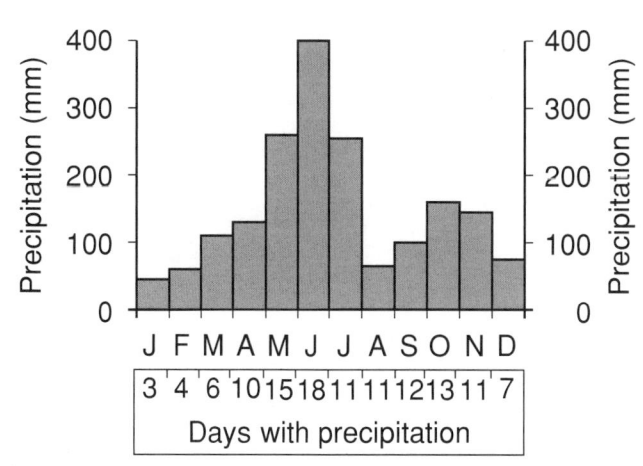

Question 2: Biosphere

(a) **Explain** fully what is meant by the term climax vegetation. 5

Study Diagram Q2.

(b) **Describe** and **give reasons** for the changes in plant types likely to be observed across the transect as you move inland from the coast. You should refer to named plant species in your answer. 13

Diagram Q2: Transect of Sand Dune System

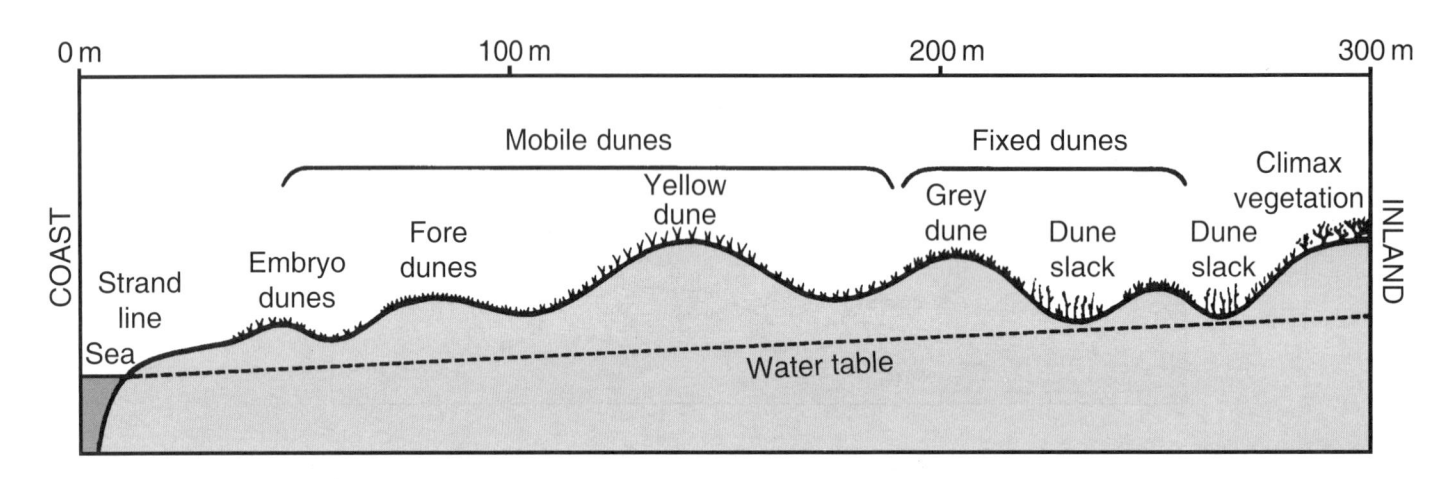

Question 3: Rural Geography

Study Map Q3.

Referring to a named area in **either** a shifting cultivation **or** an intensive peasant agricultural system:

(i) **describe** and **explain** the main features of your chosen farming landscape;   8

(ii) **describe** the recent changes that have taken place; **and**

    discuss the impact of these changes on the people and their environment.   10

Map Q3: Generalised distribution of selected agricultural systems

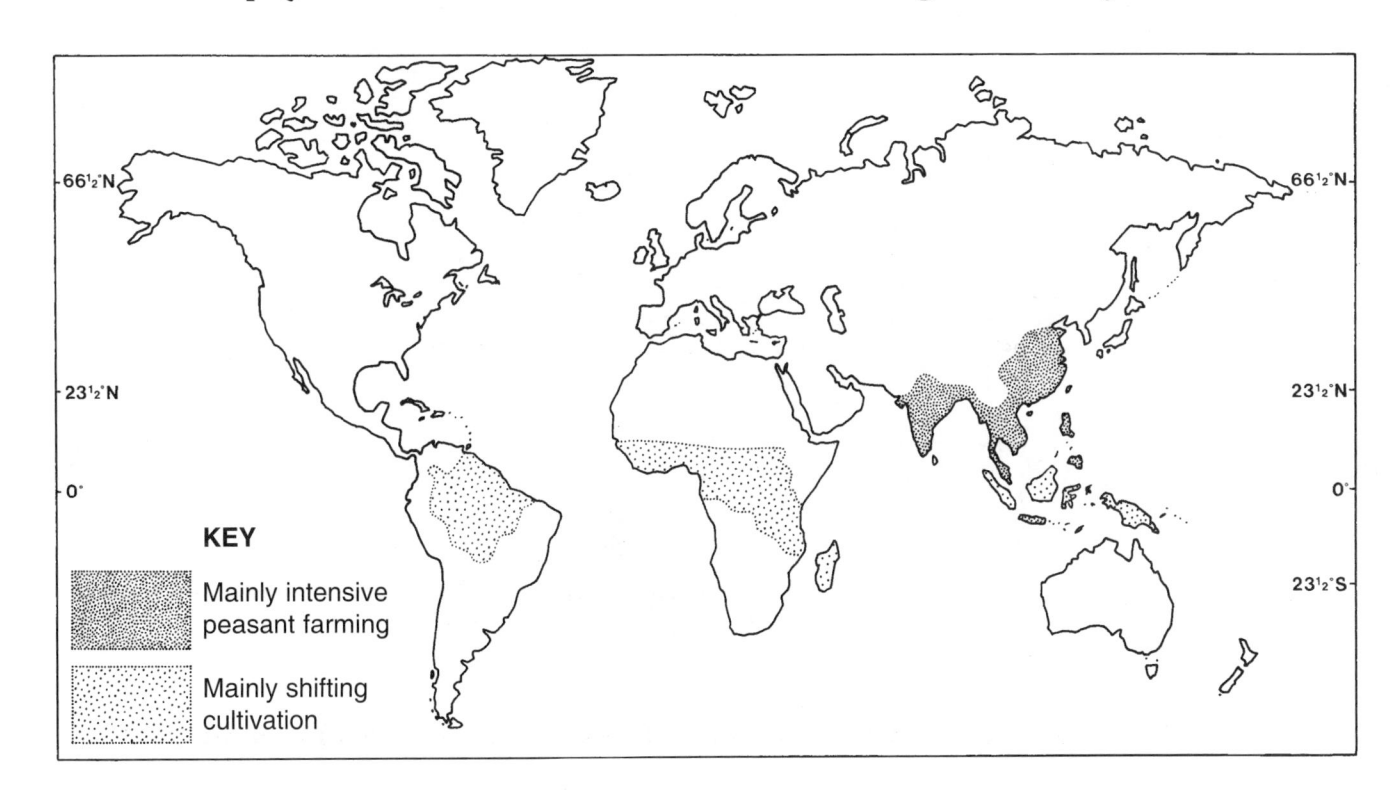

[Turn over

Question 4: Industry

Study OS Map Extract number 1882/159: Swansea (*separate item*), and Maps Q4A and Q4B.

(a) Using map evidence, **describe** and **explain** the physical **and** human factors that have encouraged industry to locate in areas A and B. **10**

(b) Study Map Q4B.

Explain ways in which the European Union **and** national government can create industrial regeneration in South Wales, **or** any other named industrial concentration in the EU which you have studied. **8**

Map Q4A: Location of industrial areas in Swansea

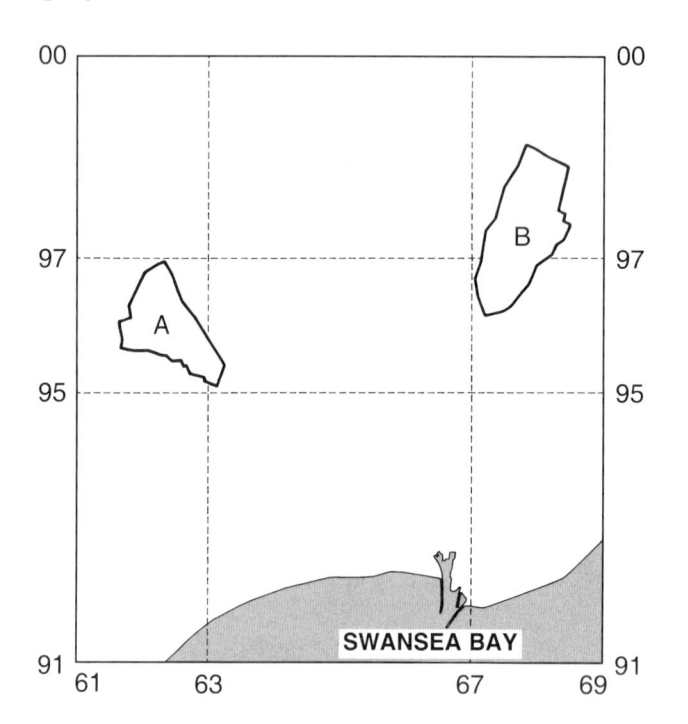

Map Q4B: South Wales—Location of new industrial developments

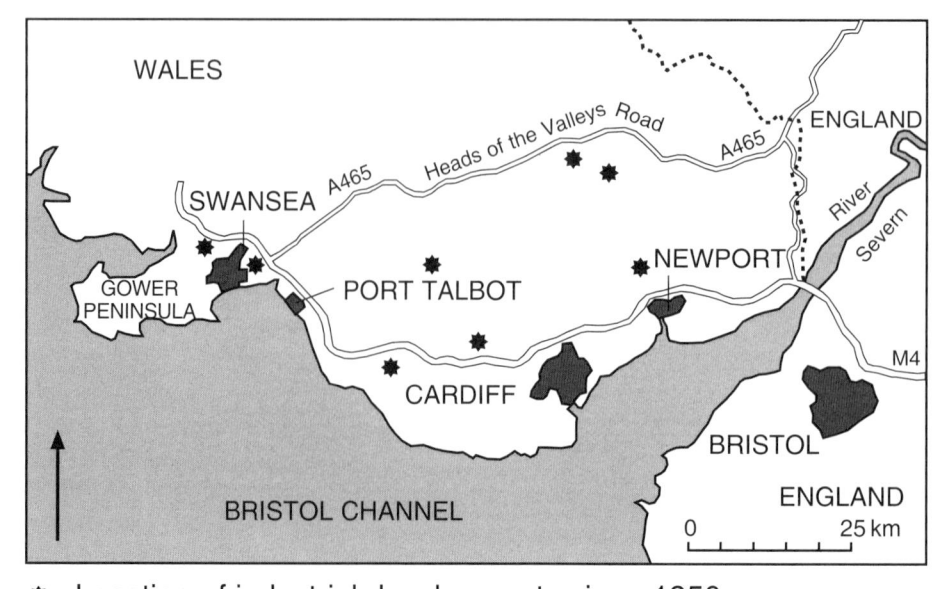

✸ Location of industrial developments since 1950

Marks

**SECTION B: Answer ONE question from this section,
ie either Question 5 or Question 6.**

Question 5: Lithosphere

Study OS Map Extract number 1882/159: Swansea (*separate item*).

(a) Using map evidence, **identify** the features of coastal erosion from GR 513851 (Oxwich Point) to GR 636871 (Mumbles Head). **6**

(b) With the aid of annotated diagrams, **explain** the formation of **one** of the erosional features described in part (*a*). **8**

[Turn over

Marks

DO NOT ANSWER THIS QUESTION IF YOU HAVE ALREADY ANSWERED QUESTION 5

Question 6: Hydrosphere

(a) Study Diagram Q6A.

"*A drainage basin is an open system with four elements—**inputs**, **storage**, **transfers** and **outputs**.*"

Describe the movement of water within a drainage basin with reference to the four elements above.

7

Diagram Q6A: A Drainage Basin

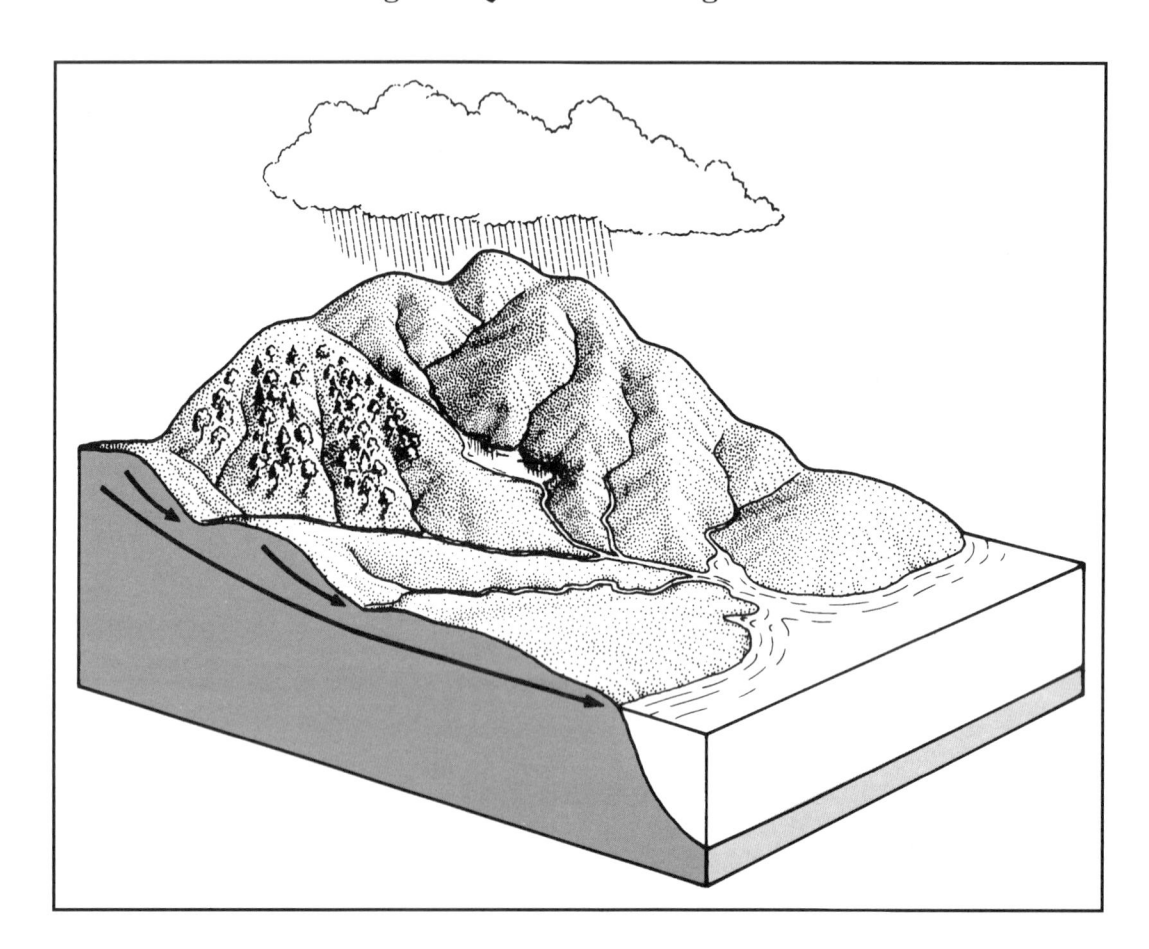

Marks

Question 6: Hydrosphere (continued)

(b) **Describe** and **explain** the changing river levels on the River Thaw at Cowbridge on 26 July 2007.   **7**

Diagram Q6B: Flood Hydrograph for the River Thaw at Cowbridge, 26 July 2007

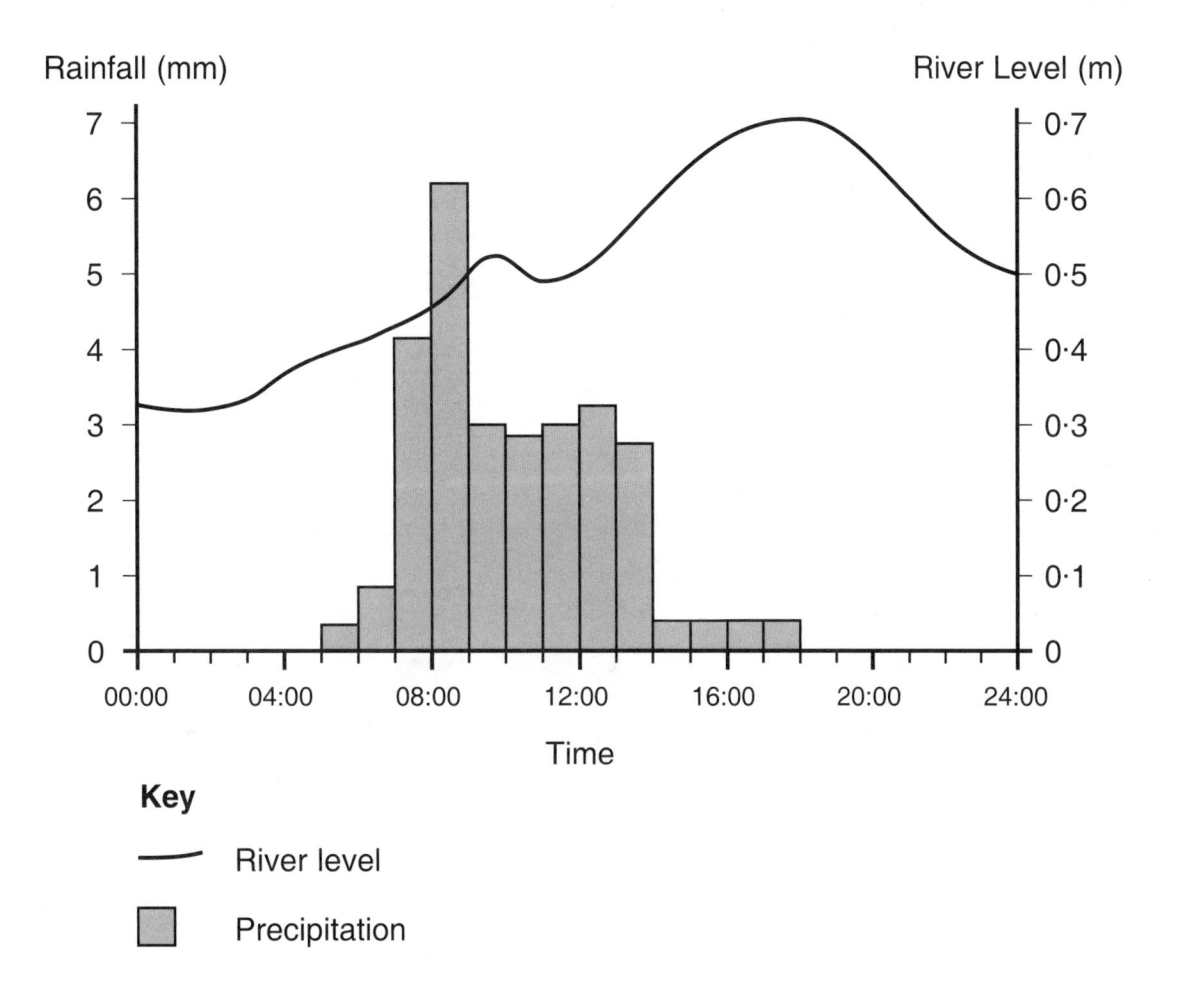

Key

— River level

▢ Precipitation

[Turn over

Marks

**SECTION C: Answer ONE question from this section,
ie either Question 7 or Question 8.**

Question 7: Urban Geography

(a) For a named city which you have studied in the Developed World, **explain** the ways in which its site **and** situation contributed to its growth. 6

(b) With reference to any Developed World City you have studied, **describe** and **explain** the land use changes in recent years in **either** the Central Business Disctrict (CBD) **or** the inner city. 8

DO NOT ANSWER THIS QUESTION IF YOU HAVE ALREADY ANSWERED QUESTION 7

Marks

Question 8: Population Geography

"*Apart from 1941, during the Second World War, the UK has carried out a census every 10 years since 1801.*"

(a) **Describe** how the UK gathers population data **between** these censuses and **explain** why it is important for countries to obtain accurate population data.   5

(b) Giving named examples, **explain** why carrying out a census may be more difficult and the results less reliable in Developing Countries than in Developed Countries such as the UK.   9

[END OF QUESTION PAPER]

X208/303

NATIONAL
QUALIFICATIONS
2011

TUESDAY, 24 MAY
10.50 AM – 12.05 PM

GEOGRAPHY
HIGHER
Paper 2
Environmental
Interactions

Answer any **two** questions.

Write the numbers of the **two** questions you have attempted in the marks grid on the back cover of your answer booklet.

The value attached to each question is shown in the margin.

Credit will be given for appropriate maps and diagrams, and for reference to named examples.

Questions should be answered in sentences.

Note The reference maps and diagrams in this paper have been printed in black only: no other colours have been used.

Marks

Question 1 (Rural Land Resources)

(a) Study Map Q1A.

Describe and **explain**, with the aid of annotated diagrams, the formation of the main features of glacial erosion in the Lake District **or** any other glaciated upland area which you have studied. **18**

(b) Study Diagram Q1.

With reference to the area around Coniston Valley, **or** any other upland area you have studied, **explain** the social and economic opportunities created by the landscape. **10**

(c) Study Map Q1B.

The Lake District and Snowdonia are both areas of outstanding glaciated scenery. **Explain** why these two National Parks attract widely differing numbers of visitors. **6**

(d) For the Lake District **or** any other upland **or** coastal area you have studied:

(i) **explain** the environmental conflicts that may occur due to an influx of visitors. (You should refer to specific named examples within your chosen area); **10**

(ii) for **one** of the conflicts explained in part (i), **describe** the solutions to this conflict and comment on their effectiveness. **6**

(50)

Map Q1A: The Lake District

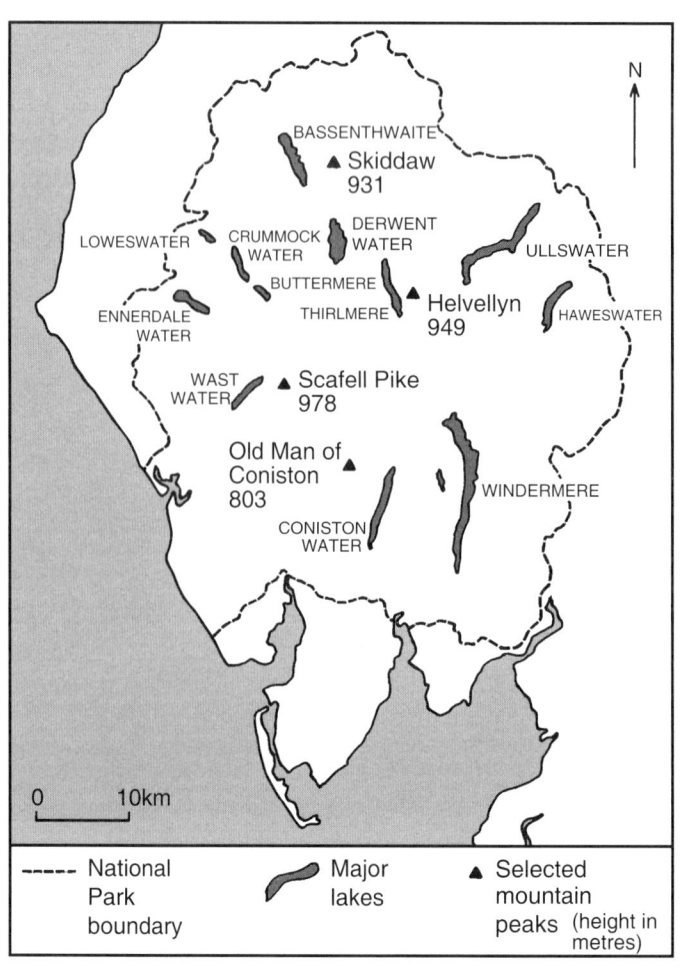

Question 1 – continued

Diagram Q1: Sketch Cross-section of the Coniston Valley in the Lake District

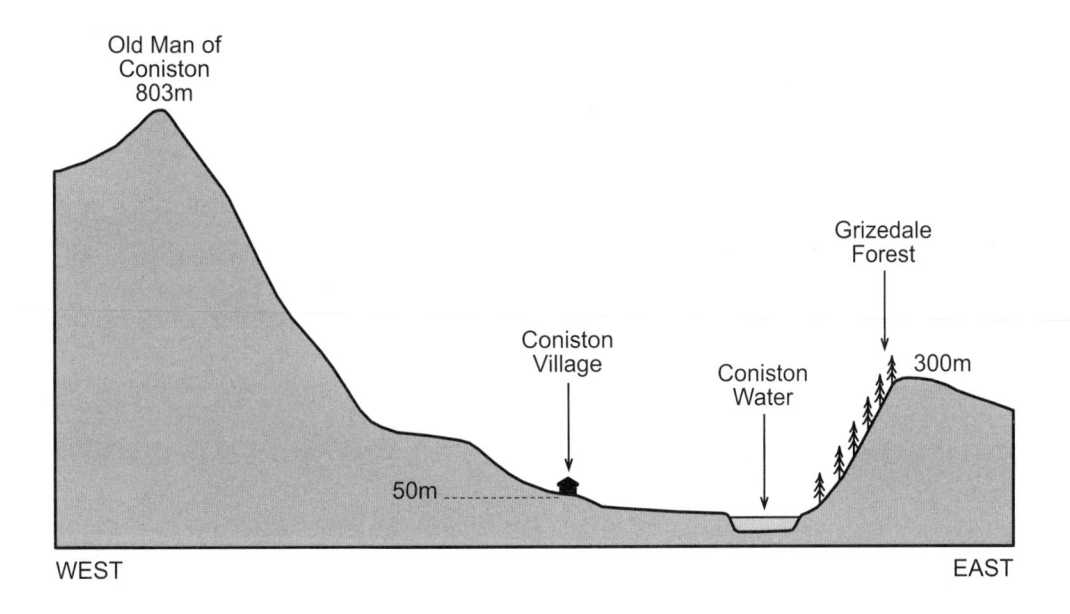

Map Q1B: Major Roads and Settlements around the Lake District and Snowdonia

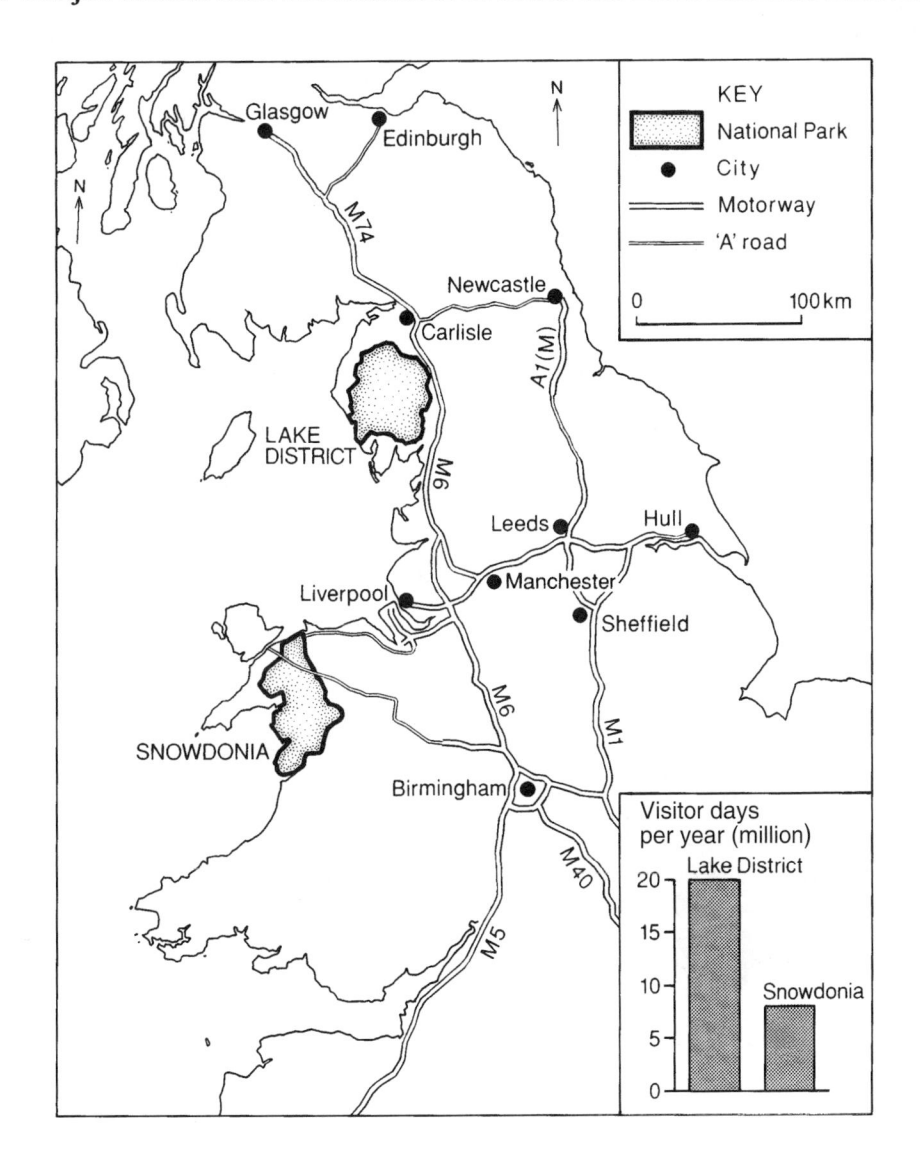

Question 2 (Rural Land Degradation)

(a) **Describe** the processes of water and wind erosion which lead to soil degradation. **12**

(b) Study Map Q2 and Diagrams Q2A, Q2B and Q2C.

Describe and **explain** why the climate of Niger has led to severe land degradation. **8**

(c) For **either** Africa north of the equator, **or** the Amazon Basin, **explain** how human activities including deforestation, overgrazing, overcultivation and any other inappropriate farming techniques have led to land degradation. **16**

(d) For named areas of North America, **describe** and **explain** soil conservation strategies that have reduced land degradation. **14**

(50)

Question 2 – continued

Map Q2: Location of Niger

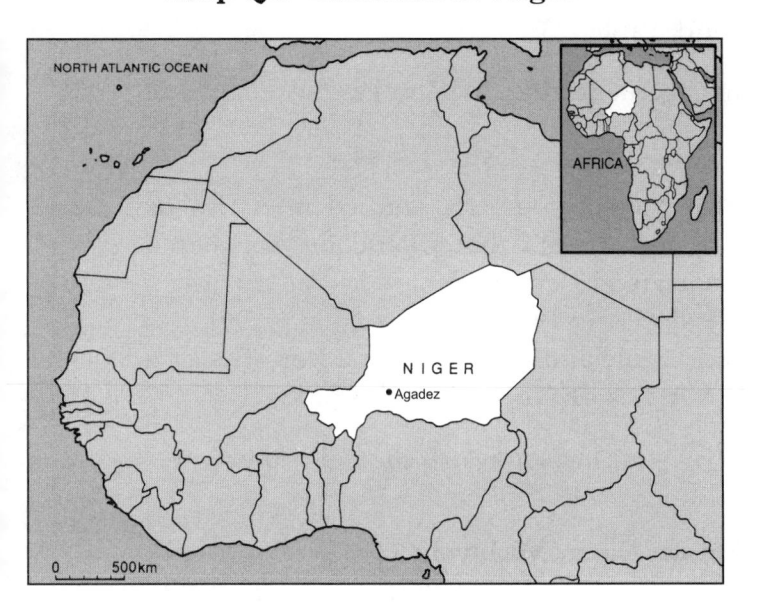

Diagram Q2A: Climate of Agadez, Niger

Key
— Temperature
▪ Precipitation

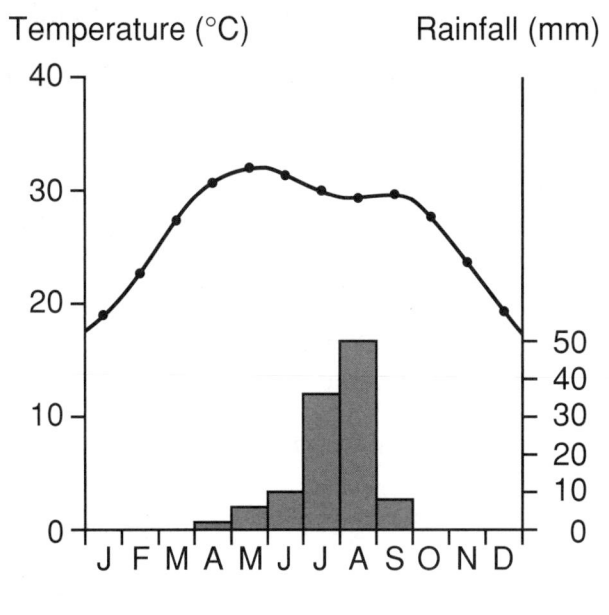

Diagram Q2B: Average daily sunshine hours, Niger

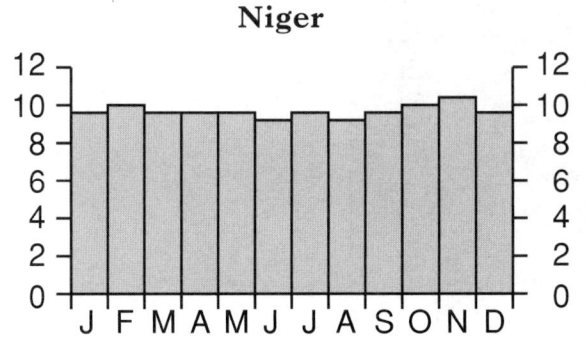

Diagram Q2C: Rainfall Variability in Niger

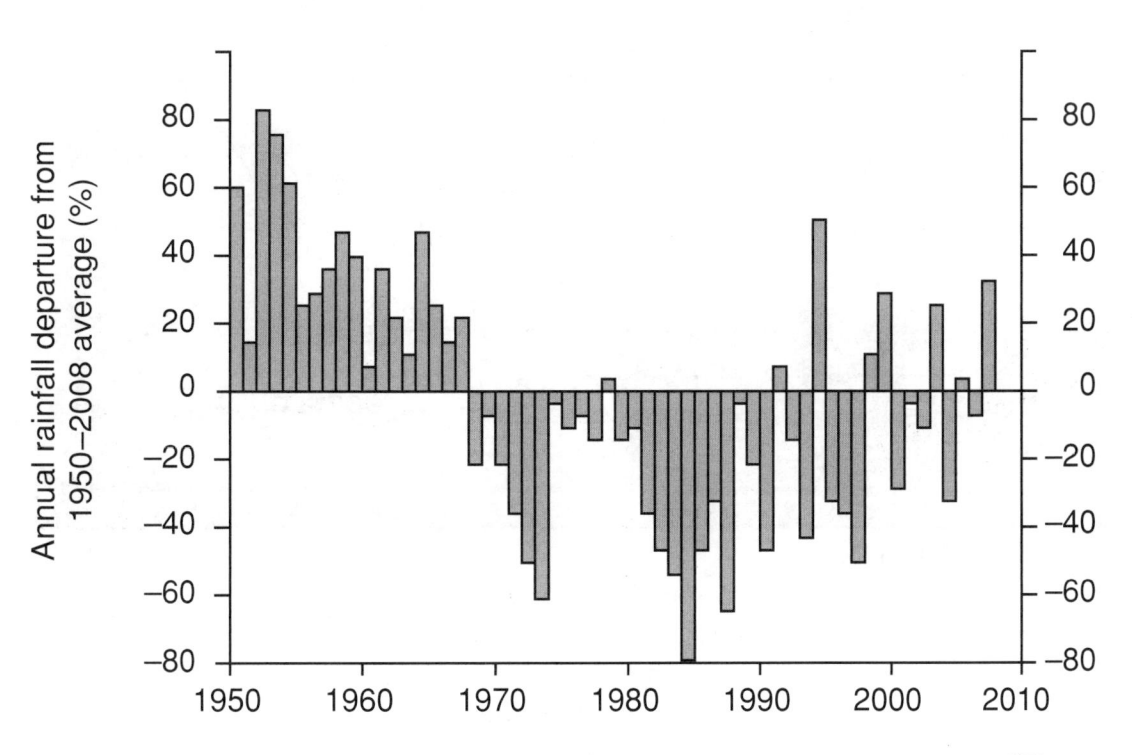

Marks

Question 3 (River Basin Management)

(a) Study Map Q3, Graphs Q3A and Q3B, and Table Q3.

Explain why there is a need for water management in the Malaysian owned area of the island of Borneo. **12**

(b) For the Bakun Dam **or** any water control project you have studied in Africa **or** North America **or** Asia, **explain** the physical factors that should be considered when selecting sites for the dam(s) and associated reservoir(s). **10**

(c) (i) **Describe** and **explain** the social, economic and environmental **benefits** of a named major water control project in Africa **or** North America **or** Asia. **16**

(ii) Comment on any **problems** caused by your chosen water control project. **12**

(50)

Map Q3: Proposed Bakun Dam, Malaysia

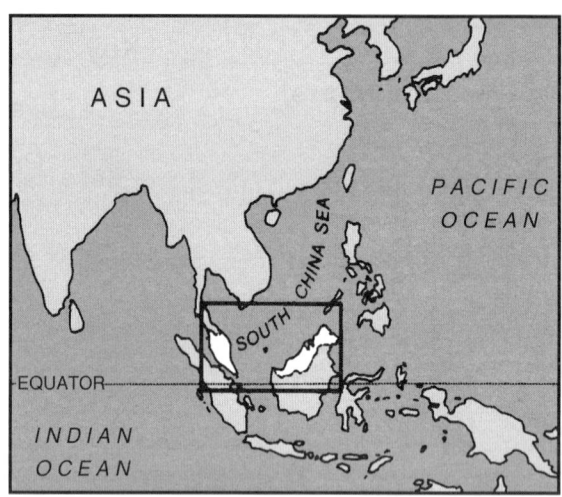

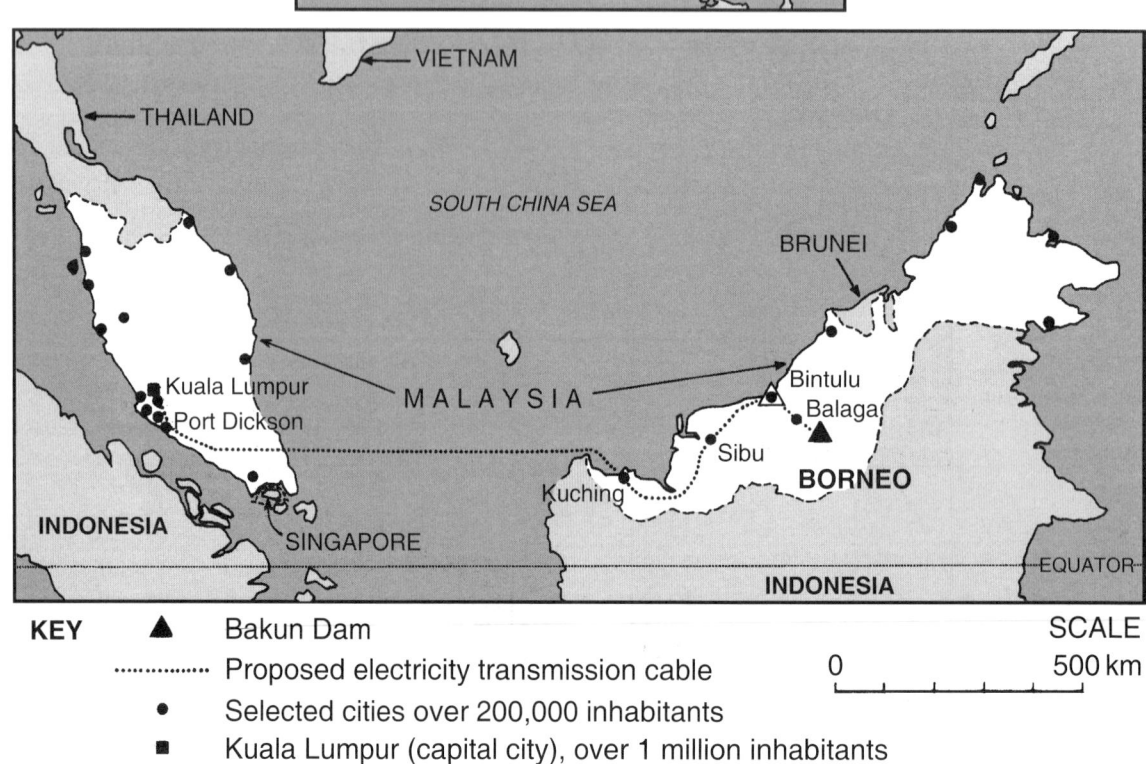

KEY
▲ Bakun Dam
⋯⋯ Proposed electricity transmission cable
• Selected cities over 200,000 inhabitants
■ Kuala Lumpur (capital city), over 1 million inhabitants
⚙ Proposed aluminium smelter (Bintulu)

SCALE 0 — 500 km

Question 3 – continued

Graph Q3A: Balaga Climate Graph

Graph Q3B: Population of Malaysia (millions)

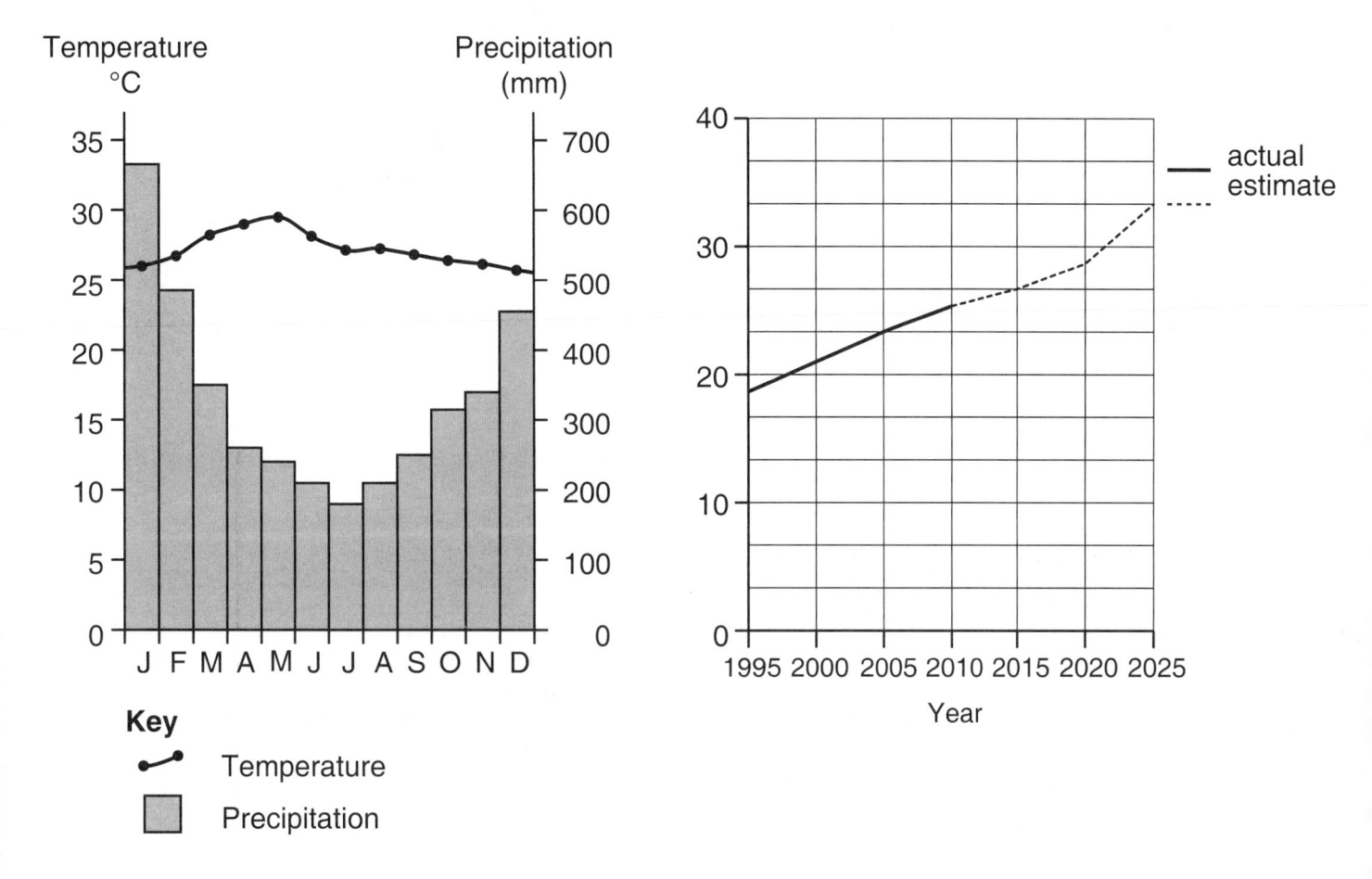

Table Q3: Indicators of Development—Malaysia

Urban population (2008)	70%
Rate of urbanisation (annual rate of change from 2005–10 estimate)	3%
Electricity production (2007)	102·9 billion kwh (kilowatt hours)
Electricity consumption (2010 estimate)	99·8 billion kwh
Electricity exports (2010 estimate)	2·3 billion kwh

[Turn over

Question 4 (Urban Change and its Management)

Marks

(a) Study Map Q4A (on *Page nine*).

Describe and **account for** the distribution of major cities in **either** the UK **or** any other **Developed World** country that you have studied.   **10**

(b) Study Map Q4B (on *Page ten*) and Table Q4 (on *Page eleven*).

> "*A Games Like No Other—Glasgow 2014 Commonwealth Games.*
>
> *Glasgow will host the 20th Commonwealth Games where 71 countries will compete in 17 sports at various venues across the city. New facilities will involve massive investment and regeneration, particularly in the east end of the city.*"

Discuss the advantages **and** disadvantages of this development for the residents of the East End of Glasgow.   **10**

(c) For Glasgow, **or** a named city you have studied in the **Developed World**, **describe** and **explain** why it suffers from traffic congestion.   **10**

(d) Many cities in the **Developing World** are experiencing rapid population growth.

With reference to a named city that you have studied in the **Developing World**:

(i) **describe** and **explain** the problems caused by this rapid growth;   **12**

(ii) **describe** the methods the residents and local authorities might use to tackle these problems.   **8**

(50)

Question 4 – continued

Map Q4A: Largest Cities in the UK

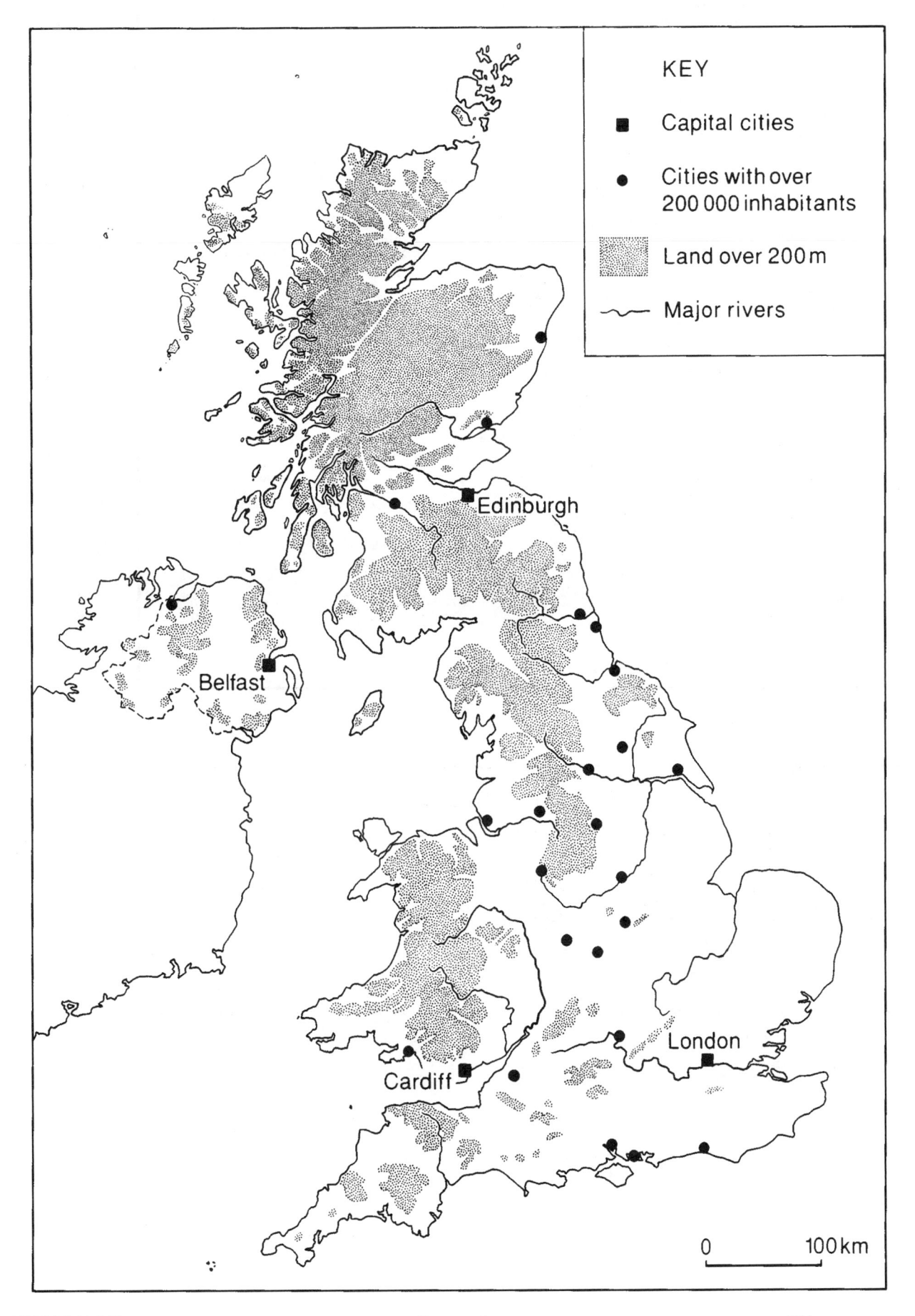

Question 4 – continued

Map Q4B: Selected Commonwealth Games Venues and Events Maps

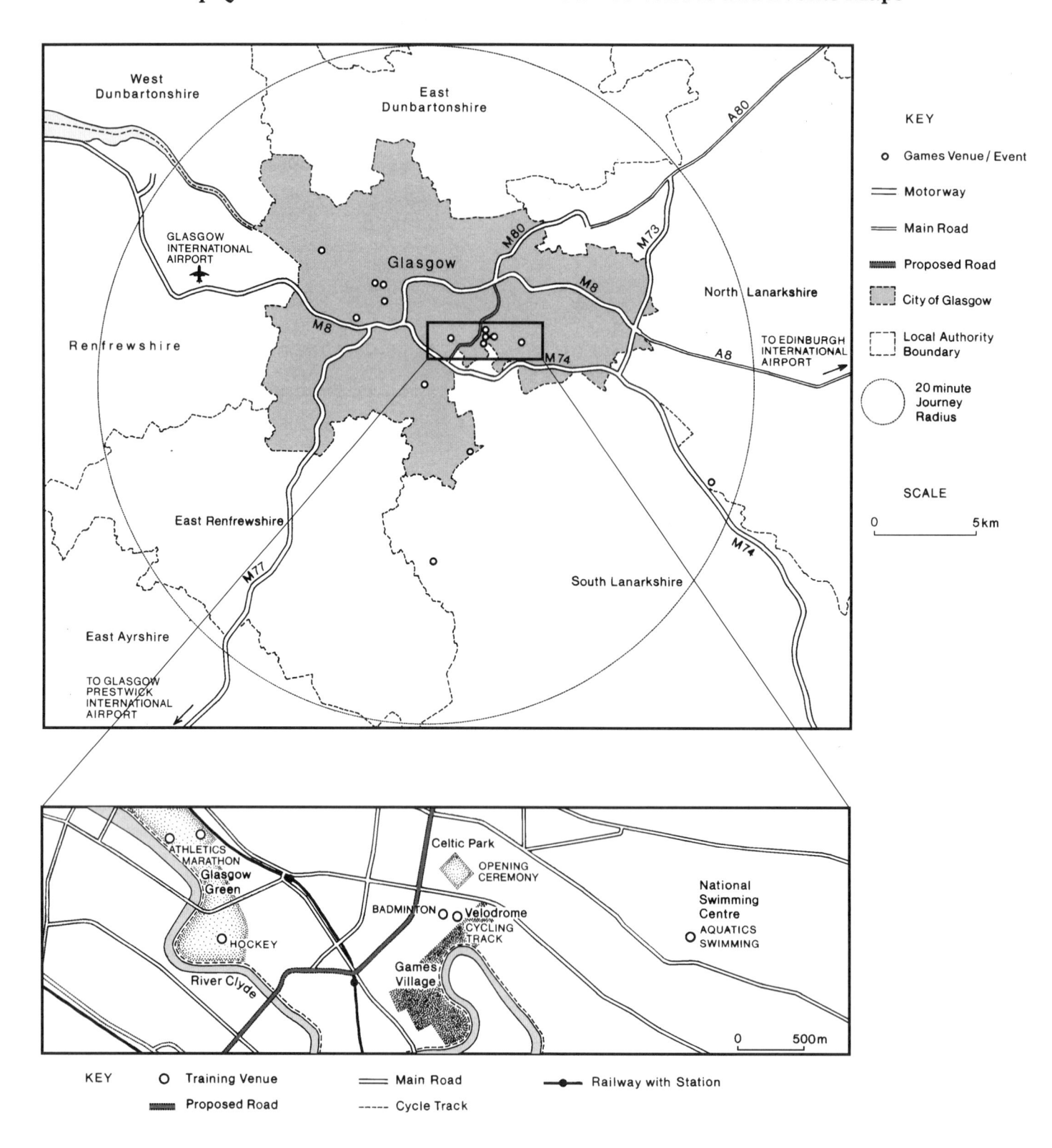

Question 4 – continued

Table Q4: Games Statistics

Public investment	£298 million (80% from the Scottish Government and 20% from Glasgow City Council)
Jobs	1000 for Glasgow, 1200 for Scotland
Net economic gain	£81 million
Volunteers	15,000 to be trained

[Turn over

Marks

Question 5 (European Regional Inequalities)

(a) Study Table Q5A.

> "The European Union in 2009 had 27 member states and a waiting list of countries who had applied to join; including Turkey, Croatia, the Former Yugoslav Republic of Macedonia and Iceland."

Suggest reasons why countries may wish to become members of the European Union. 10

(b) Study Map Q5 and Table Q5B.

Italy is often described as having a "North-South Divide". With reference to specific provinces and data provided in the table, **comment** on the accuracy of this statement. 12

(c) For Italy **or** any other country of the European Union which has marked differences in economic development between regions, **describe** the physical **and** human factors that have contributed to such regional differences. 16

(d) For the country chosen in part (c):

  (i) **discuss** ways in which less prosperous regions can receive help from their national government to overcome such problems; and

  (ii) **comment** on the effectiveness of these strategies. 12

(50)

Table Q5A: GNP per Capita data for selected EU countries

EU Country	Year of joining	GNP per capita (PPP*) 2004	GNP per capita (PPP*) 2008	% Increase in GNP per capita (PPP*)
UK	1973	$31,430	$36,130	15
France	1957	$29,460	$34,400	17
Slovakia	2004	$14,480	$21,300	47
Poland	2004	$12,730	$17,310	36

PPP* = Purchasing Power Parity

Question 5 – continued

Map Q5: Italy—GNP Per Capita by Provinces

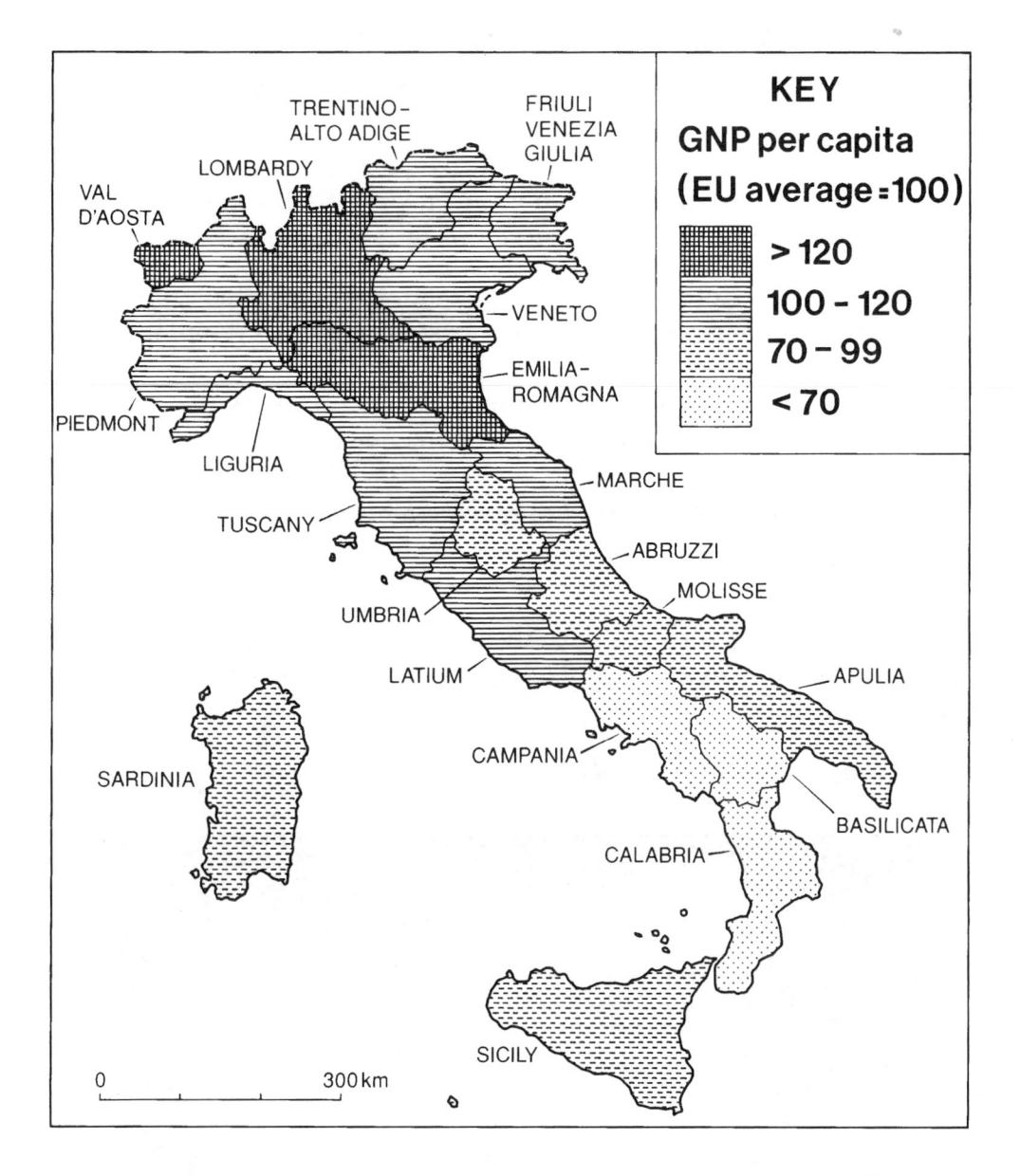

Table Q5B: Italy—Percentage of total employment in selected sectors in North and South

Employment Sector	Percentage in north	Percentage in south
Industry	78	22
Commerce	65	35
Services	59	41

[Turn over

Question 6 (Development and Health)

Marks

(a) *"Levels of wealth, health and economic development are not evenly spread within individual countries."*

Study Map Q6 and Table Q6.

(i) In what ways does the information given in the table suggest that the eight provinces of Kenya are at different levels of development? **10**

(ii) For Kenya **or** any other Developing Country that you have studied, **suggest reasons** why such regional variations exist **within** a Developing Country. **10**

(b) For malaria, **or** bilharzia, **or** cholera, **describe** the measures that can be taken to combat the disease and **explain** the varying effectiveness of these measures. **18**

(c) *"The ultimate goal of Primary Health Care is better health for all."*

(World Health Organisation)

Describe some specific Primary Health Care strategies that you have studied and **explain** why these strategies are suited to Developing Countries. **12**

(50)

Map Q6: Provinces of Kenya

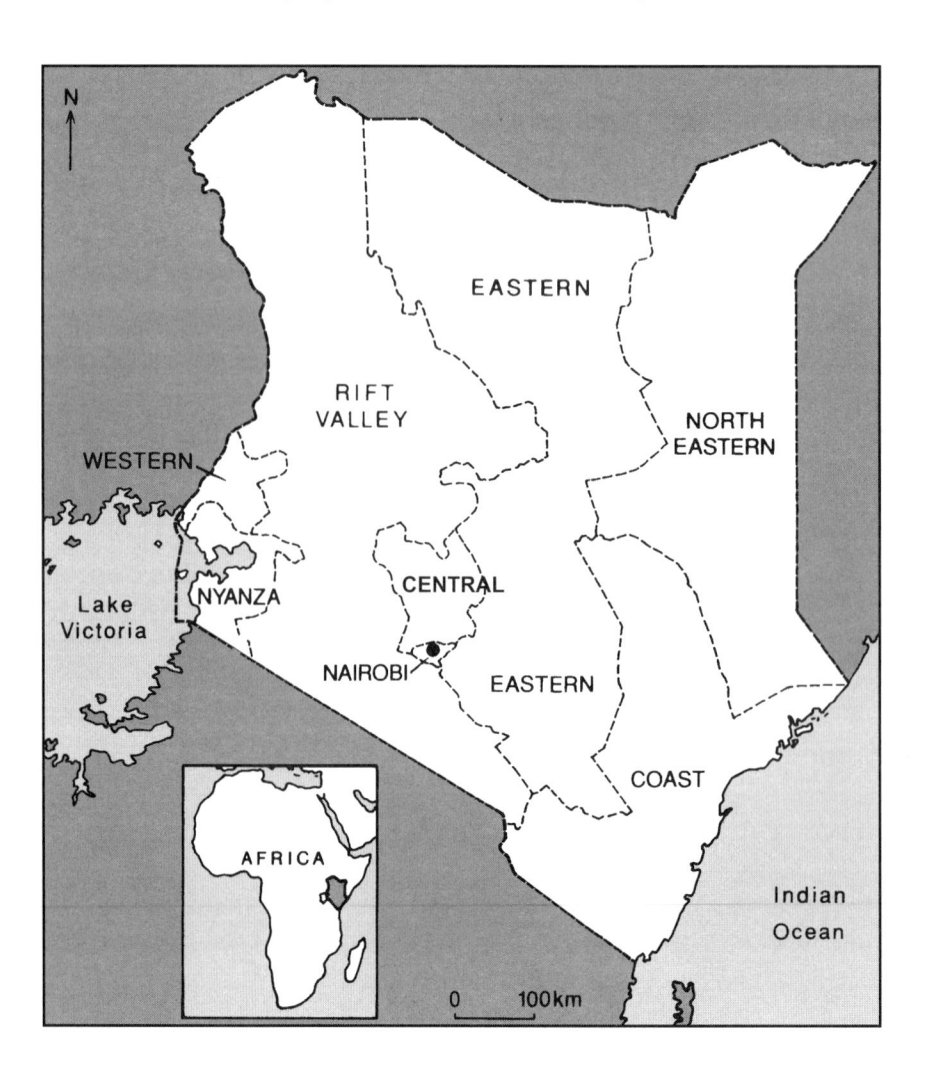

Question 6 – continued

Table Q6: Selected socio-economic indicators of development for Kenya's provinces

Province of Kenya	% Females with no education	% Males with no education	% Population below the poverty line	% Children aged 12–23 months without all vaccinations
North Eastern	87	66	64	92
Coast	38	23	58	36
Eastern	20	14	58	37
Central	12	8	31	24
Nairobi	10	8	44	40
Rift Valley	28	22	48	46
Western	18	11	61	52
Nyanza	18	10	65	64

[END OF QUESTION PAPER]

Acknowledgements

Permission has been sought from all relevant copyright holders and Bright Red Publishing is grateful for the use of the following:

A graph of global air temperatures 1855–2005 © Climatic Research Unit (2008 Paper 1 page 6);

The U.S. Census Bureau for statistics from their website (2008 Paper 1 page 7);

A photograph reproduced by permission of Buchanan Galleries (2009 Paper 1 page 10);

A photograph reproduced by permission of Braehead shopping centre. (2009 Paper 1 page 10);

Two extracts adapted from 'Trump golf resort would destroy jewel in crown of bare dunes' by Graeme Smith, taken from The Herald 19/6/08. Reproduced with permission of Herald & Times Group (2010 Paper 2 page 2);

An extract from http://www.rspb.org.uk/ourwork/conservation/sites/scotland/menie.asp © RSPB (2010 Paper 2 page 2);

Ordnance Survey © Crown Copyright. All rights reserved. Licence number 100049324.

HIGHER | ANSWER SECTION

… BrightRED ANSWER SECTION FOR

SQA HIGHER GEOGRAPHY
2007–2011

GEOGRAPHY HIGHER
PHYSICAL AND HUMAN ENVIRONMENTS
2007

Question 1 – Atmosphere

(a) Air Mass A: Tropical Continental OR cT
Origin: Over the Sahara Desert (ie large land mass in tropical latitudes)
Nature: Warm, dry, stable air

Air Mass B: Tropical Maritime OR mT
Origin: Over Atlantic Ocean (Gulf of Guinea) in tropical latitudes
Nature: Warm, moist, unstable air

(b) Description should be able to focus on the marked contrasts in rainfall amounts between a very dry north (with only 250mm per annum) and a much wetter south (where some coastal areas receive over 2000mm per annum) as shown on the map provided. Candidates could also refer to the graphs and note the variations between the three stations with Tombouctou in the north and Jos "in the middle" experiencing obvious wet and dry seasons whilst coastal Lagos in the south has a vastly greater annual rainfall total, no dry months and a "twin peak" regime.

Explanation ought to concentrate on the role of the I.T.C.Z and its associated Tropical Maritime air mass (warm, moist and unstable) and Tropical Continental air mass (warm, dry and stable). For example, Lagos – on the coast of the Gulf of Guinea – is influenced by warm moist Tm air for most of the year. This accounts for its much higher annual rainfall total. The twin rainfall peaks can be attributed to the I.T.C.Z migrating northwards and then southwards again later in the year, following the overhead sun or thermal equator. Tombouctou, in contrast, lies well to the north of the I.T.C.Z in January and is under the influence of the hot, dry Tc air from the Sahara Desert. In May/June the I.T.C.Z. moves north bringing moist Tm air and rainfall to Jos and, to a much lesser extent, Tombouctou which lies closer to its point of maximum extent.

Question 2 – Hydrosphere

(a) Descriptions may include the following points:
- meanders, ox-bow lakes, tributaries, cut offs, small pools
- broad flat floodplain, alluvium, gently sloping valley sides
- height of land and easterly flow of river could also be credited
- braiding
- islands or eyots (806533)
- river terraces.

(b) For the selected feature the explanation should include 4 well developed points, all of which could be included in a well annotated diagram.

For example for a meander:
- development of pools and riffles (differences in speed and depth)
- erosion on the outside (concave bank) of bends due to faster flow
- helicoidal flow removes material
- deposition on the inside (convex bank) of bends due to slower flow and formation of point bars
- migration of meanders downstream.

Question 3 – Population Geography

(a) Candidates ought to be able to mention such ways as census taking and compulsory registration of births and deaths. Some may include a mini or sample census (eg Britain carried out a 10% sample census in 1966) or government sponsored sample surveys (eg General Household Survey or Population Trends and Social Trends).
A reasonably detailed description of a census or census taking without reference to Civil Registration could score 3 marks, eg:
- a regular count (every 10 years in U.K.) of the population which allows comparisons to be made over time
- responsibility of each household to complete census form given out by enumerator who collects them and checks that all individuals are recorded
- a range of questions are asked designed to elicit information on individuals (age, occupation, qualifications, ethnic origins etc) and on living conditions and lifestyle (housing type, tenure, car ownership etc)
- government agencies process the data collected and make it available for interested parties
- electoral roll can update between one census and another.

(b) (i) Difficulties affecting data collection in **ELDCs** might include:
- large numbers of migrants (eg nomadic peoples such as the Tauregs or Fulani in West Africa or shifting cultivators in the rainforests of the Amazon; rural-urban migration; refugees from civil war or natural disasters) are difficult to keep track of
- significant proportions of homeless people – eg shanty or street dwellers who have no official or permanent address
- the variety of languages spoken in many countries (eg Nigeria, China, India – over 1 billion people, 15 official languages and 1650 dialects!!)
- poor communication links; difficult terrain; scattered or isolated villages
- the sheer size of many countries (eg China, India, Brazil; Indonesia with its 13 677 Islands spread over 5 100 kilometres!)
- the considerable costs involved (eg training enumerators, printing and distributing forms) impose a huge financial burden on many debt-ridden ELDCs.

(b) (ii) The reliability of data obtained may be affected by such factors as:
- higher levels of illiteracy will mean that forms are not filled in or will be completed incorrectly
- suspicion and distrust of officials in authority might lead to false answers being given as might resentment at particularly sensitive questions or those open to misrepresentation
- ethnic tensions and internal political rivalries have affected census accuracy in some countries – eg states in northern Nigeria have been known to inflate their population figures to secure increased political representation
- under-registration may occur for social and religious reasons. Reliable sources have quoted Jordanian males as stating on being asked their family size "two children and three girls"!
- China's One Child Policy must have encouraged many families to be economical with the truth! In China many female births are not registered and female infanticide has been widely reported.

Question 4 – Industrial Geography

(a) Answers will vary and will be determined by the industrial concentration chosen. For **South Wales** answers might include:
- iron ore provided the raw material for early ironworks
- coal mining saw South Wales emerge as a major industrial centre of the Industrial Revolution
- limestone was also found locally for the iron and steel works in the valleys
- large seams of coal were easily mined from the valley sides and from below the valley floors
- routeways could follow the valleys leading down to coastal ports like Cardiff
- Cardiff and Swansea developed ports around their natural bays for export of coal and manufactured goods around the world
- the British Empire provided a huge and protected market for coal and steel products from British industrial areas such as South Wales
- labour supply was provided by people migrating from rural Wales and surrounding areas of rural England.

(b) Answers will vary and will be determined by the industrial concentration chosen. For **South Wales**, steps taken to attract new industries and inward investment might include:
- creation of Enterprise Zones (Swansea, Milford Haven) and their associated benefits
- designation of Development Area status for old coal mining areas
- setting up of Welsh Development Agency (WDA) in 1976, to attract high quality investment into Wales
- Urban Development Corporation (UDC) in Cardiff and its associated benefits
- construction of New Town, Cwmbran (1949)
- relocation of specific government offices, eg DVLA in Swansea
- encouraging inward investment from abroad, eg Sony, Bosch, Lucky Goldstar
- EU (creation of EU itself provides huge European market for goods)
- joining EC (EU) opened up a huge source of funds available to outlying areas – ERDF (European Regional Development Fund), EIB (European Investment Bank), ESF (European Social Fund) etc and their associated benefits
- improved infrastructure – the Heads of the Valleys Road.

Question 5 – Lithosphere

(a) Evidence might include references to:
- deep U-shaped valley
- truncated spurs
- corries with Corrie Lochs
- flat plateau surfaces
- misfit stream
- aretes
- pyramidal peak
- hanging valley.

References might also be made to general characteristics of the landscape – bare rock surfaces and steep slopes.

(b) Glacial erosion features which could be chosen would include: U-shaped valley, arete, hanging valley, truncated spur, corrie.

In explaining the formation of a U-shaped valley, for example, candidates could refer to such points as:
- the build up of snow and ice during the Ice Age(s) caused valley glaciers to move downhill under gravity from their source in the mountains (corrie glacier) following existing (V-shaped) river valleys
- the huge weight/volume of ice combined with processes such as plucking, abrasion and rotational sliding widened and deepened these valleys
- as the valley glacier advanced it abraded the former interlocking spurs leaving truncated spurs and steepened the sides of the valley/glacial trough
- the resulting U-shape of the valley left behind when the ice melted and disappeared may vary according to rock hardness and the intensity of erosion
- the U-shape of the valleys shown in this photograph have been masked in cross-section by later deposition of moraine, scree and alluvial material on their lower slopes.

Question 6 - Biosphere

(i) For one soil, candidates should describe at least 3 out of the 4 characteristics. For example for a podzol candidates could mention:
- clearly defined horizons and iron pan
- raw acid humus layer, A – ash-grey layer, B – red/brown layer
- A – a sandy texture, B – a denser texture
- A – leached
- Iron pan in B – impedes drainage and causes waterlogging.

(ii) The following features could be included for a podzol:
- Climate – precipitation exceeds evaporation leading to leaching, and cool temperatures hinder the breakdown of plant litter.
- Natural Vegetation – coniferous forest provides a plant litter of needles and cones and develops a mor/acidic humus.
- Soil Organisms – limited soil biota and slow activity due to cool climate restricts mixing of soil and creates distinctive horizons.
- Relief and drainage – higher ground reduces temperatures, and the downward movement of water leads to leaching, and the formation of an iron pan.
- Parent material – may be fluvioglacial sands or till or acidic parent rock which determines the nature of the C horizon.

Question 7 – Rural Geography

(a) Main characteristics of shifting cultivation might include:
- clearings are made in the rainforest by cutting down and burning trees
- largest trees and some fruit-bearing trees may be left for protection/food. Some are too difficult to remove
- the 'cultivation' part refers to the practice of growing crops in the clearing, using ash from the tree-burning as fertiliser
- nutrient levels drop quickly due to heavy rains and lack of further fertiliser
- the 'shifting' part refers to the practice of moving to another clearing as the soil becomes exhausted
- system is labour intensive and only supports low densities of population
- subsistence way of life
- hunting and gathering.

(b) Impacts might include:
- area of rainforest for shifting cultivators is reducing due to cattle ranching, mineral extraction, logging etc
- indigenous population is being forced into inaccessible areas which are often less fertile, and on to reservations
- some shifting cultivators move away, (causing rural depopulation) to shanty towns surrounding large urban areas
- contact with Western cultures can bring disease
- population density increases in remaining areas, putting more strain on limited land, leading to

- fallow period decreasing with time. If fallow period too short, abandoned clearing will revert to unproductive grassland, not secondary forest
- soil fertility decreases with time and so
- output per hectare decreases with time
- soil erosion can take place – soil washed into rivers choking rivers reducing wildlife for shifting cultivators
- therefore system cannot support increasing population
- loss of wildlife habitat/biodiversity
- fewer trees/less O_2 – more CO_2 – Global Warming.

Question 8 – Urban Geography

(a) The following characteristics may be noted:
Area A
- Grid iron street pattern.
- Transport centres eg bus station, railway station.
- Bridging point.
- Important buildings eg cathedral, museum, churches, information centres.
- Congested urban landscape with many small streets and higher buildings.

Explanations may refer to accessible location, crossing point on river and early ecclesiastical function.

(b) The advantages of the residential environment may include:
Area B (suburban housing area)
- Access to motorway/A class roads for commuting eg M5 and A4538.
- Well designed road system with roundabouts for free flow and cul-de-sacs for safety and privacy.
- Pleasant environment with woodland and open spaces.
- Services available including churches and nearby schools.
- Leisure facilities including a golf course.
- Less densely populated compared to centre.
- Probably detached or semi-detached housing.

Area C (Callow End)
- Quiet commuter village.
- 5 kilometres from CBD.
- Village services including post office, pub, two churches.
- Surrounded by countryside including farmland and orchards.
- B4424 goes through village and connects with A449 into Worcester.
- Probably a variety of old and new rural housing.

GEOGRAPHY HIGHER ENVIRONMENTAL INTERACTIONS 2007

Question 1 – Rural Land Resources

(a) Descriptive points could include references to there being only two National Parks in Scotland yet a high concentration in northern England (4) and in Wales (3). Candidates could also note the absence of NPs in central and south-eastern England.

Explanations for their location ought to focus on such points as the scenic diversity and differing attractions of the various National Parks as well as on accessibility and proximity to urban (catchment) areas.

(b) (i) **Benefits** brought by an influx of tourists might include:
- new job opportunities for local people
- increased business for shops, hotels and restaurants
- more wealth generated in the local economy – "multiplier effect"
- increased property prices
- improved services (eg sports and leisure facilities) and transport links (eg upgraded roads, more frequent bus services)
- less need for young people to leave the area
- increased expenditure on conserving the amenities of the area.

(ii) Tackling the **negative** effects of tourism could include mention of:
- attempted solutions to traffic and congestion problems such as the provision of more parking facilities; promoting park and ride schemes; improving and encouraging the use of public transport; building by-passes or ringroads
- the role of visitor education through information centres, leaflets, improved signposting, promoting alternative attractions to reduce pressure on existing honeypots
- ways of restoring or preserving footpaths - eg cutting into the limestone at Malham Cove to safeguard paths, laying "terram", fencing off vulnerable areas to restrict access.

(c) Candidates ought to be able to describe and explain the formation of a reasonable range of coastal landforms.
These could include:
- headlands, cliffs, caves, arches, stacks, shore platforms (wave-cut)
- bays, beaches, spits, bars, dunes, salt marshes
- rias, fiords or sea lochs, raised beaches, fossil cliffs.

Named landforms require to be backed up by description/explanation/correct location. Authentic examples of specific features such as The Old Man of Hoy (stack) or Hurst Spit (in Hampshire) will be credited.

Question 2 – Rural Land Degradation

(a) The four main processes of erosion by water can be described as:
- rainsplash – the impact of raindrops on the surface of a soil
- sheet wash – the removal of a thin layer of surface soil which has already been disturbed by rainsplash
- rill erosion – small eroded channels, only a few centimetres deep and not permanent features, often obliterated by the next rainstorm
- gully erosion – steep sided water channels, several metres deep which can cut deeply into the soil after storms and are often permanent.

The three main processes of wind erosion can be described as:
- surface creep – the slow movement of larger (and heavier) particles across the land surface
- saltation – the bouncing along of lighter particles
- suspension – the lightest particles (dust) blown off ground for up to several hundred kilometres, dust storms.

(b) The human causes of land degradation may vary according to the locations chosen but may include:

For the **Dust Bowl**:
- use of techniques better suited to the moister east
- monoculture, especially of wheat or demanding crops (cotton), depleted the soil of moisture and nutrients
- deep ploughing of fragile soils (previously these had been held in place by natural grasslands)
- marginal land ploughed – particularly in wet years – leaving them in a fragile condition in dry years
- ploughing without due regard to slope
- farm sizes too small forcing farmers to overcrop – particularly when prices were low

For the **Tennessee Valley**:
- much of the area was cleared of its trees – this opened up the soil surface to erosion
- mining and farming also cleared the natural vegetation and led to soil erosion
- the farmers cultivated steep slopes which were ploughed up and down the slope
- overcropping had already weakened the soil
- the eroded soil was dumped in rivers and this caused them to flood, running soil further downstream

For **African north of the Equator** the following human factors might be included:

Mention might be made of:
- overgrazing, overcropping, deforestation, monoculture, burning, farming cash crops

These should be carefully explained eg:
- deforestation for firewood/building
- bush fires to clear land for farming
- in some areas (eg Tigray) small farms have led to overcropping
- in some places peasant farmers have had to farm marginal land due to the best land being used for cash crops (eg in parts of Sudan)
- the drought may have caused nomads to move into villages where the land may now be over cultivated (eg in Burkina Faso)
- rapid population growth in the countries of the Sahel has contributed to this pressure on the land

For the **Amazon Basin**:
- deforestation – for eg ranching/mineral extraction/logging/road building/poor peasant farmers – detailed accounts of these processes can be accepted
  - eg the impact of ranching: forest cleared, used for a few years until grass fails – move and clear a new stretch of forest and continue the process
  - eg the impact of charcoal smelters associated in the early years of the Carajas iron-ore mine

(c) For **Africa north of the Equator** descriptions may include:
- crop failures and the resulting malnutrition leading to famine eg Sudan, Ethiopia and much of the Sahel
- migration on a large scale – usually into shanties on the edge of the major cities

- the collapse of the nomadic way of life due to the lack of grazing and water
- many nomads forced to settle in villages – with a consequent increase in pressure on the surrounding land
- the breakdown of the settled farmer/nomad relationship in places like Yatenga province in Northern Burkina Faso

For the **Amazon** basin answers may include:
- destruction of the way of life of the indigenous people eg clashes between the Yanomami and incomers
- destruction of the formerly sustainable development eg rubber tappers and Brazil Nut collectors
- clashes between various competing groups eg the violent death of Chico Mendez allegedly at the behest of ranchers
- reduction of fallow period leading to reduced yields with obvious consequences for the dependent population
- creation of reservations for indigenous people
- increase in 'western' diseases
- increases in alcoholism amongst indigenous population

(d) Candidates must mention four methods. An example of an explanation which might receive credit is:
- shelter belts – on low lying land affected by strong winds shelter belts are rows of trees grown across the direction of the prevailing wind. They act as a barrier to slow down winds and protect the soil. The taller and more complete the barrier of trees the more effective the shelter.

Question 3 – River Basin Management

(a) For North American river basins:
- description should include reference to general patterns/numbers of rivers, and should refer to the directions of flow
- explanation should refer to the fact that drainage basins are determined by the location of the main continental watersheds and that major rivers rise in the main mountain ranges that have greater precipitation, eg the Rockies and Appalachians in North America.

Patterns within North America could distinguish between:
- west flowing rivers are fed from the western side of the continental divide. Rivers like the Columbia-Snake and the Colorado flow west in to the Pacific Ocean
- north-flowing rivers drain to the Arctic Ocean or to Hudson Bay and are fed from the Canadian Shield
- the St Lawrence system is fed from the Great Lakes areas and flows east to the Atlantic Ocean
- most of south-eastern USA is dominated by the Mississippi and its tributaries which are fed from the Rockies in the west and the Appalachians in the east and flow to the Gulf of Mexico.

(b) Explanation of the need for water management in the Narmada River Basin and in Gujarat State might include:
- reference map Q3E indicates that the Narmada River has many tributaries and the river basin has a very high drainage density leading to unpredictability of river flow which is dependant on when and how quickly snow melts in surrounding mountain areas
- rapidly increasing population in India gives increasing demand for water for domestic, power, industrial needs
- increasing demands from farmers for irrigation water to try and feed increasing population
- rainfall graph for Ahmedabad indicates seasonal nature of rainfall – extremely dry from October to May but huge monthly figures for July/August – leading to flooding and also run-off of water that could be stored and used in dry months

- temperature graph for Ahmedabad indicates hot temperatures throughout the year leading to very high evaporation rates. Monthly maximum temperatures peak at over 40°C.

(c) Political problems might include:
- river basins often cross state or international boundaries, causing difficulties in co-operation between states/countries
- sharing allocation of water rights often causes political strife
- water flow and water quality dependant on actions of upstream neighbours
- increased pollution and salinity downstream can lead to poor water quality and extra costs, eg desalination for downstream areas of river.

(d) Answers will depend upon the basin chosen. However, they might include:

Benefits:
Economic:
- more reliable water supply allows for double cropping – surplus sale
- improves navigation links
- HEP and water for industry, creating jobs and industrial expansion.

Environmental:
- flood control
- reliable seasonal water flow
- increased fresh water improves health and sanitation
- improvement in scenery?

Social:
- improved water supply
- more food available
- less disease eg cholera
- larger population sustainable
- greater availability of electricity
- opportunities for tourism and recreation.

Problems:
Economic:
- very expensive schemes
- rely on Foreign Aid, can lead to debt
- disruption of main communication links
- greater use of fertiliser increasing costs.

Environmental:
- water and industrial pollution
- silting up of reservoirs, soil erosion
- dams reduce fresh alluvial soil from renewing flood plain
- threat to wildlife habitats and historical/archaeological sites.

Social:
- local peoples forced to leave homes
- increase in water-borne disease such as Bilharzia.

Question 4 - Urban Change and Management

(a) Candidates ought to be able to pick out that 7 of the 10 largest urban areas (agglomerations) in the world today are in ELDCs whereas in 1957 the situation was reversed. Explanations for this marked change should include references to the high birth rates experienced in ELDCs over the last 50 years (compared to declining BRs in EMDCs) **and** to the various "push" and "pull" factors which account for their high amounts of rural – urban migration.

The relative decrease in city growth in EMDCs may also be attributed to such factors as planning/ environmental legislation designed to curtail outward development.

(b) (i) Social, economic and environmental problems ought to be related to the candidate's chosen city and might include:
- continued growth of shanty towns (*favelas, bustees* etc) in a range of locations in and around the city. These areas are characterised by home-made dwellings, overcrowding, inadequate water supplies, poor sanitation, disease, lack of amenities … and are often sited on fragile or unstable land liable to landslides
- unemployment/underemployment
  - growth of 'grey' economy and black market
  - drugs, crime, racketeering and prostitution
  - poor wages for unskilled jobs
  - lack of services, schools and hospitals
  - chronic traffic congestion and associated high levels of pollution.

(ii) Again, methods used to tackle the problem ought to be authentic to the candidate's chosen city! For many cities these could include mention of:
- self-help schemes (such as those in Sao Paulo) where local authorities provide basic houses made of breeze blocks and roof tiles with local residents supplying labour and digging ditches for water, sewage pipes etc and for general 'finishing off'
- money saved can then be used to provide amenities such as electricity, a clean water supply, tarred roads, a community centre, even a school, perhaps
- erecting high-rise apartment blocks, mainly in the suburbs
- building new dormitory or satellite towns to relieve the pressure on the existing metropolis (eg Cairo's Sadat City or 10th of Ramadan City).

(b) (iii) Some qualitative statement on the success/or otherwise of these schemes based on the candidate's chosen city is required. For example, the advantages of self-help schemes are that costs are kept to a minimum so that more 'basic shell' houses can be provided; they can be built in stages and working together encourages a community spirit/shared ownership.

(c) (i) Answers will, obviously, depend on the EMDC city selected but land-use conflicts on the rural-urban fringe could be attributed to such factors as:
- pressures to release land for new suburban housing estates – conflict between farming interests and developers
- building new out-of-town shopping centres/retail parks/business and science parks also makes demands on existing land users
- by-passes, ring-roads and new motorways and their associated junctions and service stations require large areas of land around cities
- recreational uses such as golf courses and country parks are still needed by the urban population and try to resist being taken over
- Conservationists/Environmentalists are against further urban sprawl and have encouraged redevelopment of brownfield or gap sites within city boundaries.

(ii) Candidates ought to be able to comment on the effectiveness of Green Belt strategies in being able to limit/control some of these demands/conflicts of interest in relation to their chosen city. Some may, in doing so, offer some comment on alternative strategies – "Green Wedges" which allow growth to take place in certain controlled directions whilst maintaining green areas close to the urban area itself have often been suggested as a compromise.

Question 5 – European Regional Inequalities

(a) Candidates should note the increasing development (from Poland-Spain-France) with increasing life expectancy, health expenditure, GDP and urban %. Explanations could include the length of membership in the EU and subsequent financial aid with development. The location of the countries within the EU and ease of trade/connection to markets could also be made relevant.

(b) A number of indicators identify regional inequalities in Poland.

Wealth shown by **GDP per capita** has province 7 (Mazovia) well ahead (31 100 zloty) with provinces 3 (Lublin) and 9 (Sub-Carpathia) the lowest (under 15 000 zloty). Wealth shown by % **private vehicles** – only provinces 7 (Mazovia), 12 (Silesia) and 15 (Greater Poland) are in double figures.

Electricity production % shows two provinces 5 (Lodz), and 12 (Silesia) with 20% and 7 (Mazovia) next with 13%, three provinces have less than 1% (4, 10 and 14).
Province 7 (Mazovia), including the capital city Warsaw, is shown to be the most developed province in Poland using these indicators
Candidates could compare provinces and/or statistics but should use some form of comparative statement(s) covering all three indicators.

(c) Candidates should be able to illustrate their answer with reference to specific named locations.

(i) Candidates responses will vary according to the region studied but the description may include some/all of the following:
Social problems
- high levels of youth/long term unemployment
- chronic crime, drug abuse, cycle of deprivation, welfare dependency culture
- organised corruption
- depopulation, ageing population.

Economic problems
- low investment, high % of working population in primary/declining industry
- low skills base in working population
- poor infrastructure
- long term poverty.

Candidates could refer to both physical and human factors to account for these problems.
Physical factors eg more mountainous environments and harsher climates.

Human factors eg remoteness/isolation and poorer communications; distance from markets; decline of traditional industries/raw materials.

(ii) Measures taken by national governments may include:
- government incentives: Regional Development status, Enterprise Zone status, capital allowances, training grants, and assistance with labour costs, rent free arrangements
- specific assistance to old industrial areas eg coal mining (in UK/Belgium) or long established industrial areas eg San Sebastian area in Northern Spain
- government intervention: relocation of specific government departments (eg Civil Service jobs to Glasgow)
- government directions: state owned firms directed to invest in specific areas (eg former Britoil to Aberdeen or car firms such as Fiat to Southern Italy)
- government pressure/support: major multinational conglomerates (eg VAG/SEAT location in southern Spain).

Measures taken by European agencies may include:
- European Regional Development Fund (ERDF) – provides a wide range of direct and indirect assistance to encourage firms to move to disadvantaged areas eg improvements to local infrastructure, job creating investments, local development projects and aid for small firms
- European Investment Bank (EIB) – concentrates on providing loans for businesses setting up in disadvantaged areas
- European Social Fund (ESF) – promotes the return of the unemployed and disadvantaged groups to the workforce, mainly by financing training measures and systems of recruitment assistance
- Cohesion Fund – this is a special fund designed to assist the least prosperous countries of the Union (the 10 new Member States as well as Ireland (until 2003), Greece, Portugal and Spain). This fund co-finances major projects involving the environment and trans-European transportation networks.

The effectiveness of the measures will be dependant upon the country studied and the region chosen. Some measures of success in the UK for example with Enterprise Zone status, and the provision of training grants. Charleroi in Belgium has also shown evidence of regrowth thanks to robust government intervention. Both national and European measures must be covered.

Question 6 – Development and Health

(a) **Social** indicators could include:
- infant mortality rate per thousand
- number of persons per doctor
- number of cars/TVs/telephones per 1 000 people.

Economic indicators could include:
- gross domestic product per capita
- average income per head
- percentage of working population employed in, say, the Primary sector
- energy consumption per capita.

(b) Differences in the levels of development between Economically Less Developed Countries (ELDCs) may be due to:
- mineral reserves eg Saudi Arabia and similarly positioned Middle East countries have vast reserves of oil. They also have stable (if despotic) government regimes/monarchies that leads to the generation of huge wealth. This wealth can 'trickle down' to a wide sector of the population. Other countries may have no reserves of minerals in demand by the EMDCs (Economically More Developed Countries)
- political instability eg many have unstable regimes or are suffering from border wars and/or civil wars eg Sudan, Indonesia, Rwanda
- colonial links eg some Caribbean countries receive support from western countries because of their former colonial ties
- strategic locations eg South Korea and many Central American countries receive additional support
- encouragement of entrepreneurial skills and the ability to attract in major world companies eg by offering an educated, resourceful and relatively cheap work force (South Korea) and/or incentives eg 10 years rent free factory sites in Vietnam
- natural disasters eg Bangladesh (cyclones & floods), Indonesia (tsunami), Niger (recurring drought and associated famines) will limit progress.

(c) Candidates responses will vary according to the ELDC they choose. Differences could include; North v South; poorer rural and relatively rich urban areas; within cities ie richer suburbs and shanty areas; subsistence farming v commercial farming areas; tourist 'honeypots' and remote interiors; areas rich in minerals and those with rural depopulation.

(d) (i) Answers will depend on the disease chosen but for cholera answers might include:

Physical factors:
- High temperatures.
- Estuaries and marine coastal areas with sources of shellfish.
- Earthquakes/tsunamis or floods which might disrupt water and sewage services.

Human factors:
- Poor sanitation.
- Use of unclean water.
- Sewage mixing with water supplies.
- Contaminated shellfish.
- Poorly cooked food, fruit and vegetables washed in contaminated water.
- Contaminated ice used in drinks.
- Overcrowding causing pressure on water supplies and sanitation.
- War and famine putting pressure on water supplies and sanitation.

(ii) Strategies might include:
- scrupulous care over cooking and washing food
- scrupulous hygiene
- good education re food and hygiene – boil it, peel it or forget it!
- vaccinations
- treated water eg boiling water, adding iodine or chlorine, using bottled water
- Primary Health Care
- oral rehydration salts treatment (80 – 90% of patients) or intravenous fluids
- antibiotics
- avoid buying from street vendors eg fruit or ice/ice cream
- fly control and regulations for dumping waste
- max 1 for named examples of drugs or insecticides.

(iii) The benefits of controlling the disease on a developing country might include:
- saving money on health, medicine, doctors, drugs etc
- reduction in the national debt
- healthier workforce and increased productivity
- longer life expectancy and decreased infant mortality rates
- scarce financial resources could be spent on other areas such as education or housing
- more tourists/foreign investment may be attracted if there was less risk of disease – leading to more job opportunities, foreign currency earnings, increased prosperity.

GEOGRAPHY HIGHER PHYSICAL AND HUMAN ENVIRONMENTS 2008

SECTION A

Question 1 – Lithosphere

(a) **Map evidence of Coastal Features**
  (i) *Areas A and B (Erosion)*
  - The shape of the coastline - both Area A (chalk) and Area B (limestone) are made of harder rocks, less easily eroded and therefore they jut out into the sea as headlands.
  - Name evidence eg Natural Arch, The Pinnacles, Old Harry, Tilly Whim Caves, Peveril Point, indicating a headland/cave/arch/stack coastline.
  - Off the headland at 054825 are small islands, indicating stacks or stumps.
  - Symbol evidence with cliffs and steep slopes all round Durlston Bay and along south coast of area B.
  - Wave-cut platform off Peveril Point at 042786.

  (ii) *Area C (Deposition)*
  - The shape of the coastline - Area C (sands and clays) of softer rocks, more easily eroded and therefore cut back into as a bay.
  - Name evidence eg Dunes at 038855, Studland Bay, Little Sea (lagoon).
  - Symbol evidence with sand symbol along Studland Bay and sand, heath, marsh and lagoon in 0284, 0384, landward of Studland Bay.

(b) **Formation of coastal features**
  Stack
  - Candidates should refer to the processes of coastal erosion in their answer ie hydraulic action, corrasion (abrasion), solution (corrosion) and attrition.
  - The terminology in the question 'various stages in the formation of' should encourage candidates to start their answer with a line of weakness in a headland and progress this through cave and arch formation to stack formation.

  Sand Bar
  - Candidates should refer to the processes of coastal deposition and transportation in their answer ie wave movement up and down beaches and longshore drift.
  - The terminology in the question 'various stages in the formation of' should encourage candidates to start their answer with wave movement and progress this through longshore drift and spit to bar formation.

Question 2 – Biosphere

Strandline: *Sea Sandwort; Sea Rocket; Saltwort*
- These are all salt-tolerant (halophitic) species and can withstand the desiccating effects of the sand and wind. Some can even withstand periodic immersion in sea water. The presence of these plants leads to the further deposition of sand and the establishment of Sand Couch Grass and Lyme Grass. The high pH figures can be attributed to a high concentration of shell fragments. ($CaCO_3$).

Embryo Dune: *Sea/Sand Couch; Lyme Grass*
- These dune pioneer species grow side (lateral) roots and underground stems (rhizomes) which bind the sand together. These grassy plants too, can tolerate occasional immersion in sea water. Some species found on the strandline are, of course, also found on the embryo dunes (Sea Rocket and Sea Twitch).

Fore Dune: *Sea Bindweed; Sea Holly; Sand Sedge; Marram Grass*
- A slightly higher humus content (from decayed plants), and lower salt content (further from the sea) allows these species to further stabilise the dune and allow the establishment of Marram Grass which becomes a key plant in the build up of the dune.

Yellow Dune: *Marram Grass; Sand Fescue; Sand Sedge; Sea Bindweed; Ragwort*
- Both the humus content and the acidity of the soil have increased at this location. Marram can align itself with the prevailing wind to reduce moisture loss; it can also survive being buried by the shifting sand of the dune. In fact, as sand deposition increases the Marram responds by more rapid rhizome growth (up to a staggering 1 metre a year). It is xerophytic, and so is better able to survive the dry conditions of the dune than other plants. This allows it to become the dominant species on the Yellow Dune. It also has long roots which help to bind deposited sand and anchor it into the dunes as well as access water supplies some distance below.

Grey Dunes and Slacks: *Sand Sedge; Sand Fescue; Bird's Foot Trefoil; Heather; Sea Buckthorn; Grey Lichens (eg Cladonia species)*
- As a result of an increase in organic content (humus), greater shelter and a damper soil a wider range of plants can thrive here. Marram dies back (contributing humus) to be replaced by other Grasses, Sand Fescue and Sand Sedge. As a result of leaching and the build up of humus the soil is considerably more acidic again supporting a wider plant community. In the wetter slacks, close to the water table, several water-loving (hydrophytic) species may survive; various reeds and rushes; cotton grass; alders and small willow trees.

Climax
- In some areas heathland may dominate with a range of heathers being prominent, whilst in the shell rich areas of the Western Isles machairs may develop. Eventually trees such as birch, pine or spruce could establish a foothold.

Question 3 – Population Geography

(a) Changes in Total Population
- *Stage 1*
  - Total population fluctuates but population growth is low, as high Death Rate (DR) due to wars, famine and epidemics is balanced by high Birth Rate (BR).
- *Stage 2*
  - Rapid population growth as DR falls due to medical advances eg vaccinations, improved water supply and sanitation and marked decrease in Infant Mortality Rate (IMR).
  - BR remains high due to lack of contraception and family planning, children seen as an 'economic asset' and parents wanting many children as an 'insurance policy' for being looked after in old age until IMR is seen to fall.
- *Stage 3*
  - Despite rapidly falling BR, continued rapid population growth as DR continues to fall, with continued improvements in medicine and standards of living.
  - BR falls due to the awareness of family planning and that smaller families are needed with decrease in IMR; children now seen as an 'economic liability'.
  - Population growth levels off at end of stage 3 as BR and DR reach similar low levels.

(b) Stage 5 problems caused by low birth rate/ declining (and ageing) population
- Need to maintain an active population large enough to allow levels of taxation to remain constant or raise retirement age.
- Need to ensure there are no future shortages in workforce – need to recruit immigrant labour/ease access for asylum seekers. This can lead to civil unrest/ethnic tension.
- Need to sustain demand for particular products or services eg toys, schools, maternity hospitals, which if affected could lead to higher levels of unemployment.
- Ageing population gives increased cost of pension provision and unpopular decisions for government about how pensions should be funded.

Question 4 – Urban Geography

(a) Answers will vary according to the city studied but may include:

Site
- Flat land.
- Inside a large river meander.
- Early functions eg religious, defensive, trading site.
- Bridging point on river.

Situation
- Easily accessible to major settlements.
- Accessible to ports.
- Major route focus.
- Accessible to airports.

(b) Answers will vary according to the city studied but may include:
- Ring road around city centre.
- Use of roundabouts to improve flow.
- Pedestrianised areas in the centre.
- Park and ride schemes.
- One way systems.
- Parking restrictions and fines.
- Multi-storey car parks.
- Bus lanes/improved public transport.

SECTION B

Question 5 – Atmosphere

Physical factors:
- Solar activity: variations in solar energy and sun spot activity.
- Changes in the earth's orbit and tilt: Croll-Milankovitch Cycles, 'wobble and roll'.
- Volcanic eruptions: dust particles reduce temperatures by shielding the Earth from incoming insolation.
- Ice cap/sheet melting: reduction in Albedo effect.
- Changing oceanic circulation - El Nino/La Nina.

Human factors:
- Carbon dioxide: from burning fossil fuels - road transport, power stations, heating systems, cement production and from deforestation (particularly in the rainforests) and peat bog reclamation/development (particularly in Ireland and Scotland for wind farms).
- CFCs: from aerosols, air conditioning systems, refrigerators, polystyrene packaging etc
- Methane: from rice paddies, animal dung, belching cows - even flatulent termites!
- Nitrous oxides: from vehicle exhausts and power stations.

- Sulphate aerosol particles and aircraft contrails: global 'dimming' - increase in cloud formation increases reflection/absorption in the atmosphere and therefore cooling.
- Atomic bomb - linked to cooling.

Question 6 – Hydrosphere

(*a*) Diagrams should include the key processes within the global hydrological cycle:
- Precipitation.
- Evaporation/Transpiration.
- Condensation.
- Infiltration/Run-off/Melting.
- Storage ie ice, ground water, ocean.

(*b*) Differences could include:

Interception
- Rural - there is a longer 'lag time' between the rainfall and peak discharge in the rural hydrograph because vegetation (eg woodland) will intercept precipitation and store/absorb it thus preventing the water reaching the soil/ground water/river quickly.
- Urban - concrete/tarmac/buildings will channel precipitation to gutter/drains and straight into the sewer/river system with a correspondingly shorter 'lag time'.

Surface run-off
- The rising limb is much steeper in the urban hydrograph because natural water courses will overflow and drain into marshy areas/fields on the flood plains in times of flood whereas urban water courses will be lined and embanked to contain and speed up the flow of water.

Storage
- The falling limb on the urban hydrograph is much steeper due to the lack of infiltration/percolation/underground storage of water. In rural areas water will continue to flow into the river many hours after the rainstorm through underground and through flow via the soil and rocks. The return of the river to the base flow will therefore be much slower with a more gentle falling limb.

SECTION C

Question 7 – Rural Geography

(i) System matches to associated diagram

Intensive Peasant Farming
- Traditionally high labour input although this is beginning to decline as poorer farmers are forced off the land.
- Small capital input although this is increasing with amalgamation of uneconomic holdings and increased use of machinery.
- Small parcels of land but this is also increasing with amalgamations.
- Large output due to intensive nature of system with maximum use of land available.

Commercial Arable Farming
- Labour force small and declining with increased use of large machines as agribusiness takes over from family farms.
- High input of capital, used for machinery, irrigation, pesticides, fertilisers and infrastructure.
- Very large areas of land required for effective operation of large farm machinery.

- Large output is related to huge area involved rather than particularly high output per hectare.

Shifting Cultivation
- Small labour force due to subsistence nature of system which is unable to support a large population.
- Very low input of capital related to subsistence nature of system.
- Land area is large as cultivators move from area to area within forest.
- Very low output as only a tiny proportion of land area required is cultivated at any one time.

(ii) Changes in farming practices:

Intensive Peasant Farming
- The use of mini-tractors (rotavators) and small mechanised rice-harvesters instead of draught animals less labour-intensive.
- The widespread adoption of higher yielding/faster maturing new varieties of rice - the impact of the 'Green Revolution'.
- Amalgamation of small uneconomic holdings/consolidation of fragmented fields as a result of land reform.
- The formation of farming co-operatives has provided farmers with several benefits, easier access to machinery, cheaper credit facilities, bulk purchasing of inputs and improved marketing opportunities.
- Greater use of modern pesticides and fertilisers, coupled with new seed varieties and improved irrigation systems has meant a shift from subsistence farming towards more commercial farming with small surpluses for sale.

Commercial Arable Farming
- Amalgamation of farm holdings as family farms are taken over by agribusiness.
- Part-time farming and co-operatives have increased, with greater use of contractors for harvesting.
- Diversification of crops eg away from wheat to eg sunflowers as crop/market demands change.
- Increase in organic farming reflecting change in market demand.
- Increased use of more carefully managed irrigation systems.
- Increased awareness of avoidance of soil erosion by employing a variety of soil conservation methods.

Shifting Cultivation
- Area of rainforest for shifting cultivators has been constantly reducing due to logging, cattle ranching, mineral extraction, HEP etc
- Population increases put more pressure on limited land.
- More indigenous people have been forced into more inaccessible (and often less fertile) areas, on to reservations or have moved to city living.
- Cultivated areas have to be returned to sooner after shorter fallow period leading to overworked soils which suffer a decline in fertility.
- Previously cultivated areas abandoned due to soil erosion and silting up of rivers.

Question 8 - Industrial Geography

(*a*) Old Industrial Landscape:
- *Air:* pollution from coal burning factories, railway engines and houses. Smoke, dust, soot and smog discolouring buildings and affecting the health of the people living in close proximity to their work. (High incidence of lung diseases like bronchitis).

- *Water:* untreated effluent and sewage from the buildings entering the local streams with resulting damage to ecosystems.
- *Land:* subsidence and land slippage due to mining and waste tipping with little control/health and safety laws. (Aberfan).
- *Buildings:* closely packed communities of housing/factories/transport. High population densities and overcrowding leading to poor environmental quality - little greenery or open space. Tall, brick factories with metal gratings over windows, chimneys etc – poor visual quality.

(b) New Industrial Landscape location factors could include both physical and human factors:

Physical factors:
- Flat land for easy construction of large low factory buildings.
- Room for future expansion; space for car parking and storage.

Human factors:
- Proximity to markets ie large urban areas.
- On edge of urban area for cheaper land costs/rent and proximity to labour force in housing estates/suburbs.
- Close to other modern industries that supply components or provide a market.
- Close to motorways/main roads-easy access for deliveries/work force.
- Close to airports ie for foreign executives.
- Close to ports for export/import.
- Close to universities ie source of highly skilled employees and possible partners in research projects.
- Government and EU incentives/grants.

GEOGRAPHY HIGHER ENVIRONMENTAL INTERACTIONS 2008

SECTION 1

Question 1 – Rural Land Resources

(a) Candidates should mention both surface and underground features such as:
- limestone pavements
- sink/swallow/potholes
- dolines/shakeholes
- disappearing/resurgent streams
- dry valleys
- gorges
- scars and scree
- caverns
- stalagmites, stalactites and pillars.

Appropriate explanations should be provided for the formation of the features, eg for a limestone pavement:
- glacial erosion has scraped clear the overlying soil and exposed the limestone
- joints in the limestone have been formed as a result of pressure release
- these joints/lines of weakness are enlarged due to the action of chemical weathering: rainfall is a dilute acid (carbonic acid) and the limestone (calcium carbonate) can therefore be dissolved by this rainwater. This leads to deep gaps (grykes) and raised blocks (clints).

(b) (i) The opportunities provided by the landscape could include:
- tourism, recreation, conservation
- farming, forestry
- water supply
- quarrying.

Each land use should be linked to a landscape feature, and at least two land uses are required.

(ii) Explanations of problems and conflicts will depend on areas studied but will generally include:
- air and noise pollution from tourist traffic
- traffic congestion
- erosion of footpaths
- damage to fences and walls
- litter
- water pollution
- visual pollution from new complexes, car parks etc
- noise and air pollution from quarrying
- heavy traffic polluting, congesting and damaging rural roads.

(c) If candidates selected National Parks as their conservation strategy answers may include the following points:
- the landscape is protected by legislation
- National Parks were set up to 'conserve and enhance natural beauty'
- National Parks will undertake to promote understanding and enjoyment of areas
- National Parks have a strong voice in issues like developments seeking planning permission
- National Parks will have information centres, newspapers etc which will provide information about conservation strategies.

Question 2 – Rural Land Degradation

(a) Candidate descriptions may include:
- high temperatures throughout the year
- rainfall peaking during "summer" months/drought in "winter"

- northern areas of Burkina Faso having less rainfall than southern areas
- long term patterns of rainfall showing a decrease over a 30 year period
- fluctuating annual rainfall can also be assumed.

Candidates should explain why these patterns lead to land degradation. Points might include desertification, erosion from rain and wind, over-cultivation during wet periods and impact of drought on vegetation.

(b) Answers will depend on the area chosen but may include:
- over-cultivation depleting soils of nutrients
- monoculture removing specific nutrients
- deep ploughing of fragile soils making them susceptible to erosion
- irrigation leading to salinisation
- deforestation of slopes to increase farmland leading to gully erosion
- removal of shelter belts leading to susceptibility to wind erosion
- lack of organic fertilisers used
- cultivation of marginal land leading to problems during drier years
- overgrazing of pasture leading to loss of vegetation cover.

(c) If candidates select the Sahel as their chosen case study answers may include:
- crop failure and resulting malnutrition leading to famine
- disease and illness can become endemic
- migration on a large scale, often into shanties on the edge of cities
- collapse of traditional way of life eg nomadism
- many nomads forced to settle in villages, causing pressure on surrounding land
- conflict within countries as people move and re-settle
- countries rely on international aid
- a cycle of poverty develops.

(d) Soil conservation methods might include:
- crop rotation
- diversification of farming types
- keeping land under grass or fallow
- trash farming/stubble mulching
- replanting shelter belts
- strip cultivation and intercropping
- increased irrigation
- soil banks by keeping soils under grass rather than ploughing
- diversification by farmers into recreation
- contour ploughing
- terracing
- use of natural fertilisers
- gully repair
- re-afforestation of slopes and marginal land.

Question 3 – River Basin Management

(a) Candidates may mention a range of reasons to explain the need for water management including:
- very low rainfall in Egypt (desert conditions)
- flood control
- regulating flow and storage of water
- power supply for expanding cities and industry
- water for industrial purposes
- water for agricultural irrigation as food demands increase
- drinking water for increasing population
- maintaining a navigable river.

(b) Physical factors might include:
- solid foundations for a dam
- consideration of earthquake zones/fault lines
- narrow cross section to reduce dam length
- large, deep valley to flood behind the dam
- lack of permeability in rock below reservoir
- sufficient water supply from catchment area
- low evaporation rates
- impact on the hydrological cycle.

(c) Answers will depend upon the basin chosen. However, some suggestion are outlined below:

Benefits	Adverse Consequences
Social:	**Social:**
• greater population can be sustained with increased food supply • less disease and poor health due to better water supply and more food being available • recreational opportunities • more widespread availability of electricity eg 16% of Egypt's total in 1998 • floods in 1998 and 1999 and droughts from 1979 to 1987 avoided in Egypt	• forced removal of people from valley sites eg 90 000 Nubians from the Aswan High dam/reservoir site • increased incidence of water borne diseases such as Bilharzia in irrigation channels
Economic	**Economic:**
• improved farming outputs with possible surplus for sale • HEP - industrial development creating job opportunities. (Aswan paid off more than 20 times the cost of the dam in 10 years - it produces more than 10 billion kilowatt hours of electricity every year) • water for industry • navigation opportunities	• huge costs of new schemes eg Aswan cost 1 billion US $ • dependence on foreign aid/finance in the case of ELDC's - consequent debt • more money required for fertilisers eg 1 million tonnes now used in Egypt • reduction by up to half the sardine and shrimp stock of the delta (although now recovering) • possible dislocation of communication links
Environmental:	**Environmental:**
• increased fresh water supply improves sanitation and health • scenic improvement?	• water pollution and industrial pollution • loss of alluvial supplies to flood plain/delta • silting up of reservoirs • flooding of archaeological/historical sites eg UNESCO provided 40 million US $ to rescue Abu Simbel and 19 other monuments

(d) Political problems will depend on the chosen river basin but may include reference to:
- water control/dependence on neighbours upstream eg 86% of the Nile's annual flow comes from Ethiopia
- pollution levels across borders eg 17% of Lake Nasser lies in Sudan
- shared costs with limited benefits
- complex legislation over appropriate water shares and how these are determined eg in the 1959 Nile Water Agreement Egypt and Sudan ignored Ethiopia and in the 1960's Egypt blocked the funding for 29 irrigation/HEP projects in Ethiopia
- reduction in water flows in some areas
- difficulty of predicting further demands.

SECTION 2

Question 4 – Urban Change and its Management

(a) (i) Description could focus on:
- the increasing urbanisation of the world's population in **both** More Developed and Less Developed regions
- the more rapid increase in urbanisation since 1970 in Less Developed regions
- the differing rates of increase between named areas in both More Developed and Less Developed regions (eg the urban population share has increased more significantly between 1970 and 1994 in Europe than in already highly urbanised North America). Similarly rates of urbanisation in recent and forthcoming decades are faster in Africa than in more urbanised Latin America.

(ii) Reasons for the differences could include:
- some explanations for higher birth rates experienced in ELDCs compared to fairly stable or even declining birth rates in EMDCs
- rural "push" and urban "pull" factors
- legislation such as Green Belt policies designed to curtail outward expansion of urban agglomerations in EMDCs.

(b) (i) Social, economic and environmental problems ought to be related to the candidate's chosen city and might include:
- overcrowding, inadequate water supplies, poor sanitation, lack of amenities, high incidence of disease... Lack of services, schools and hospitals
- unemployment/underemployment
    - growth of 'grey' economy and black market
    - drugs, crime, racketeering and prostitution
    - poor wages for unskilled jobs
- unsightly, homemade dwellings made out of a range of discarded materials. These are often sited on fragile or unstable land liable to landslides. Chronic traffic congestion and associated levels of pollution.

(ii) Again, methods used to tackle the problem ought to be authentic to the candidate's chosen city! For many cities these could include mention of:
- self-help schemes (such as those in Sao Paulo) where local authorities provide basic houses made of breeze blocks and roof tiles with local residents supplying labour and digging ditches for water, sewage pipes etc and for general "finishing off" (eg joinery work)
- money saved can then be used to provide amenities such as electricity, a clean water supply, tarred roads, a community centre, perhaps a school
- erecting high-rise apartment blocks, mainly in suburbs
- building new dormitory or satellite towns to relieve pressure on the existing metropolis (eg Cairo's Sadat City or 10th of Ramadan City.)

(c) Answers will obviously be dependent on the change chosen and the EMDC city studied. For **housing** change in inner city **Glasgow**, for example:

(i) Factors responsible for the change would be such points as age and poor condition of many of the city's tenements which could be described in some detail.

(ii) Ways in which Glasgow's inner city housing areas have been changed could include:
- wholesale demolition of the worst of the tenements
- the creation of Comprehensive Development Areas
- 'decanting' residents to peripheral housing estates, New Towns and further afield - Overspill Policy
- more recently the emphasis has been on urban renewal/regeneration of older properties (eg sandblasting exteriors, new roofs and windows, security doors, modernising interiors – new kitchens and bathrooms etc) and the building of new more up-market privately owned flats.

Candidates could clearly enhance answers with reference to named developments/ locations in their chosen city.

(iii) Comments on the success or otherwise of such strategies might refer to the reluctance of some older people to move; the problems associated with huge housing estates (eg lack of amenities such as a pub on the corner of every street, high crime rates, distance from workplaces) and high-rise flats (eg shoddy build quality, dampness, noise, vandalism). Many of the flats built to re-house former tenement dwellers have, in recent years, had to undergo extensive repair and, in some cases, have even been knocked down. Others may comment on the relative success of New Towns and the popularity of refurbished traditional homes.

Question 5 – European Regional Inequalities

(a) New Eastern Periphery reasons for lack of prosperity might include:
- less favoured area climatically - very cold continental winters
- more remote from major political centres of Europe - London, Strasbourg, Paris, Brussels
- less well served by transportation links - away from major motorway networks, away from 'hub' airports like Schiphol, London and Paris airports and away from major container ports like Europort with worldwide trade links
- some areas are distant from Euro-Core - miss out on economies of scale and benefits from agglomeration factors
- countries with former centrally planned economies making the transition to market economies with persisting legacy in some cases of high unemployment or underemployment and high percentage with relatively poor living standards.

(b) Indicators show the following:
- Euro-Core countries have significantly higher GDP/capita than either pre-2004 or new Periphery countries. They also have a very different industrial structure with a very small percentage in Primary Industry and a high percentage in the Services sector.
- pre-2004 Periphery countries are in a 'middle' position with regard to GDP/capita but their IMR is very similar to the Core countries. Industrial structure shows a much greater reliance on Primary Industry in Greece and Portugal compared to Core countries. Greece's structure is very similar to that of Estonia
- new Periphery countries have slightly higher IMRs indicating weaker health provision. GDP/capita is only slightly more than half that of Germany and the Netherlands, indicating a much lower wealth base.

Industrial structure varies with Slovakia's low Primary Industry figure more similar to that of Core countries

(c) The UK's regional inequalities stem from a combination of the physical differences between the higher and steeper land to the north and west of the UK compared with the lower and more gently sloping land to the south and east coupled with the remoteness of the north-west compared to the proximity of the south-east to the 'core' of the EU. Candidates may justifiably stress the positive and negative aspects of different regions.
- Physical factors might mention advantages/ problems such as relief, rock types, climate and water supply, soil fertility and erosion.
- Human factors might mention decline of traditional heavy industries, growth areas of new lighter industries and hi-tech industries, out-migration from north and differences in accessibility related to communications and remoteness.

(d) National government measures taken will vary according to whether candidate chooses a less developed region within UK, Italy or Belgium (or other).
UK national government help could include eg for South Wales:
- creation of Enterprise Zones (Swansea, Milford Haven) - giving credit for associated benefits
- the setting up of the Welsh Development Agency to attract high quality investment in Wales
- Cardiff as an example of an Urban Development Corporation
- development area status within the UK
- the construction of new towns such as Cwmbran
- the Heads of the Valleys Road as an example of improved infrastructure
- the DVLA to Swansea as an example of relocation of government offices.

EU measures would include:
- ERDF (European Regional Development Fund) - encourages firms to move to disadvantaged areas - loans, grants, improvements to local infrastructure
- EIB (European Investment Bank) - loans for factory modernisation etc
- ESF - (European Social Fund) - grants to improve job opportunities - retraining for redundant miners - assistance with relocation
- Problem/peripheral areas may be given Objective 1 or Objective 2 status which makes them eligible for funding packages/support for a set time.

Question 6 - Development and Health

(a) (i) Economic indicators could include:
- Gross Domestic Product per capita
- Average Annual Income per capita
- Percentage of working population employed in, say, the Primary sector.

Social indicators could include:
- adult literacy rates (%)
- average life expectancy at birth
- infant mortality rates per 1 000 live births
- number of cars/TV sets/telephones etc per 1 000 people.
(*Strictly speaking, the Human Development Index is a social welfare index which is calculated by giving each country a score based on: adult literacy rates, average life expectancy and average income per person adjusted to reflect local spending power.*)

(ii) Answers ought to be able to refer to the likes of:
- oil-rich countries such as Saudi Arabia, Brunei, the UAE or to relatively well-off countries like Malaysia which are able to export primary products such as tropical hardwoods, rubber, palm oil and tin as opposed to poorer nations such as Burkina Faso or Chad which lack significant resources
- Newly Industrialising Countries (NICs) eg China, South Korea, Taiwan are able to earn substantial amounts from steel-making, shipbuilding, car manufacturing, electrical goods, toys, clothing etc. They have been able to benefit from their population's entrepreneurial skills and low labour costs
- some countries such as Brazil and Malaysia have both resources and growing manufacturing industries
- the expansion of tourism has helped to improve living standards/create new job opportunities in countries like Thailand, Jamaica, Kenya, Malaysia, Sri Lanka and earns valuable foreign currency
- many countries are afflicted by recurring natural disasters which restrict development/hamper progress eg
  - drought in sub Saharan Africa (Mali, Chad, Burkina Faso...)
  - floods/cyclones in Bangladesh
  - hurricanes in the Caribbean
  - tsunamis in Sri Lanka, Indonesia
- political instability - eg recent disruptive civil wars in places such as Sudan/Rwanda/Somalia/ Liberia/Sierra Leone or larger-scale conflicts in Iraq or Afghanistan have also had a negative impact. Widespread corruption and mismanagement have accounted for the marked decline of Zimbabwe's economy and are a continuing problem in many other African nations.

(iii) Answers will, obviously, depend on the ELDC chosen but for Brazil could include:
- the South East is much more prosperous than other regions due to the concentration of industry and commerce in the "Golden Triangle" of Sao Paulo, Rio de Janeiro and Belo Horizonte.
  - this area has the best transport system in Brazil, the greatest number of services, and has benefited most from Government help
  - coffee growing has long been carried out on the rich *terra rossa* soils around Sao Paulo producing job opportunities and creating wealth for the area and the national economy
  - Rio de Janeiro - until 1960 the capital of Brazil, had the advantages of a good natural harbour which encouraged trade, immigration, industry, and more recently, tourism.
- the North East, in contrast, is handicapped by more 'negative' factors such as periodic droughts, fewer mineral resources and a shortage of energy supplies all of which have encouraged outwards migration
- the North (Amazonia) suffers from its more peripheral location, its inhospitable (Rainforest) climate, poor soils, dense vegetation and inaccessibility. Not surprisingly, it is the poorest of Brazil's five main regions. Until recently, there was also a lack of government investment and much of the region has lost out on basic services such as health, education and electricity.

In addition to explaining the sorts of marked socio-economic regional variations which exist in a huge and diverse country such as Brazil, candidates may also comment on the marked differences in living standards which exist between relatively wealthy and better provided for urban areas compared to poorer more isolated rural areas and to the contrasts that can be found *within* urban areas - eg hillside *favelas* such as Rocinho in Rio versus the prosperous apartments overlooking Copacabana Beach.

(b) (i) Answers will depend on the disease chosen but for Malaria might include:

Environmental factors:
- suitable breeding habitat for the female anopheles mosquito - areas of stagnant water such as irrigation channels, water barrels, padi fields, puddles etc
- hot wet climates such as those experienced in the tropical rainforests or monsoon areas of the world
- temperatures of between 15°C and 40°C
- areas of shade in which the mosquito can digest human blood.

Human factors:
- nearby settlements to provide a 'blood reservoir'
- areas of bad sanitation, poor irrigation or drainage
- exposure of bare skin.

(ii) Measures taken to combat Malaria can include:

Trying to eradicate the mosquitoes:
- insecticides eg DDT - however this is environmentally harmful - impacts on the food chain and is supposed to be banned as a result. In addition the mosquitoes build up a resistance to chemical insecticides through time and they become less effective
- newer insecticides such as Malathion - these are oil-based and so more expensive/difficult for ELDCs to afford - also stains walls and has an unpleasant smell - so not popular!
- mustard seed 'bombing' - become wet and sticky and drag mosquito larvae under the water drowning them
- egg-white sprayed on water - suffocates larvae by clogging up their breathing tubes (as with mustard seeds - wasteful, costly and fairly impractical)
- BTI bacteria grown in coconuts. Fermented coconuts are, after a few days, broken open and thrown into mosquito-infested ponds. The larvae eat the bacteria and have their stomach linings destroyed! Cheap, environmentally friendly and 2/3 coconuts will control a typical pond for up to 45 days.
- larvae eating fish - effective and a useful additional source of protein in people's diets
- drainage of swamps - requires much effort - not always practicable in the Tropics.

Treating those suffering from Malaria:
- drugs:
  - Chloroquin - easy to use/cheap but mosquitoes are developing resistance to it
  - Larium - powerful, offers greater protection but can have harmful side effects
  - Malarone - fairly new drug - said to be 98% effective - few side effects but very expensive
- vaccines:
  - still being developed/not yet in widespread use (eg Dr Manuel Pattaroya's in Colombia)
- education programmes:
  - insect repellent eg Autan
  - cover skin at dusk when mosquitoes are most ravenous!
  - sleep under treated mosquito nets - fairly cheap

- Quinghaosu - extracted from plant - used as a traditional cure in China for centuries - now in pill form - easy to take - may be the long awaited breakthrough.

No one solution has been found. A combination of strategies/control methods, combined with increasing public awareness/education programmes (eg WHO's 'Roll Back Malaria' - a global campaign aimed at halving the number of malaria cases by 2010) will be needed just to keep malaria in check. Some progress may be made thanks to the millions which the Bill and Melinda Gates Foundation has set aside for research into a cure.

GEOGRAPHY HIGHER
PHYSICAL AND HUMAN ENVIRONMENTS
2009

SECTION A

Question 1 – Hydrosphere

(a) Descriptions could include:
- meanders, waterfalls (eg 979664 or 999634), tributaries/confluences, braiding/eyots/islands (982653), river cliffs (Loup Scar at 030618).
- variations in the width of the valley eg broad, flat flood plain – approx 700 metres wide – in 9768 whereas in other sections such as in square 0162 the river is flowing in more of a gorge.
- references to the height of the land, steepness of the valley sides, direction of flow.
- accept direction S/SE.
- reference to speed only if explained or names.

(b) Answers should be based on the concept of differential erosion. The following points may be made:
- harder rock overlying softer rock.
- softer rock is eroded more easily by the force of running water.
- eventually, the softer rock is worn away.
- this causes undercutting so there is nothing to support the harder rock above which collapses.
- some of this shattered rock will be swirled around by the river (especially at times of spate) and helps to excavate a deep plunge pool at the base of the waterfall.
- this process is repeated over a long period of time causing the waterfall to gradually retreat upstream usually leaving a steep-sided gorge.

Question 2 – Biosphere

(a) A podzol soil profile should be drawn and annotations could include:

Associated vegetation is coniferous forest or heather moorland.

Thin black humus layer divided between layers of leaf litter (L), fermentation (F) and mor humus (H) with a pH of 3.5-4. Plants have shallow, spreading roots.

Ash-grey upper A horizon with sandy texture.

Zone of eluviation of humus, Fe and Al minerals and clay. Well-defined horizons – few soil biota to mix soil due to cold climate.

Iron pan develops in upper B horizon, impeding drainage and causing waterlogging.

Zone of illuviation with accumulation of clay, and Fe and Al oxides.

B horizon is reddish-brown with denser texture. Precipitation exceeds evaporation, giving downward leaching.

C horizon is parent material, generally weathered rock or glacial or fluvio-glacial material.

(b) The following features could be included for a brown earth soil:
- deciduous forest vegetation provides deep leaf litter, which is broken down rapidly in mild/warm climate.
- soil colour varies from black humus to dark brown in A horizon to lighter brown in B horizon where humus content is less obvious.
- soil biota break down leaf litter producing a mildly acidic mull humus. They also ensure the mixing of the soil, aerating it and preventing the formation of distinct layers within the soil.
- texture is loamy and well-aerated in A horizon but lighter in the B horizon.
- precipitation slightly exceeds evaporation, giving downward leaching of the most soluble minerals and the possibility of an iron pan forming, impeding drainage.
- trees have roots which penetrate deep into the soil, ensuring the recycling of minerals back to the vegetation.

Question 3 – Rural Geography

(a) The following points should be developed:
- Bangladesh is an ELDC with a very low GDP per capita.
- a high percentage of Bangladesh's low GDP is derived from its intensive peasant farming.
- a very high percentage of Bangladesh's working population are intensive peasant farmers.
- due to the intensive nature of farming in Bangladesh the amount of fertiliser used is high, although still less than Canada.
- Canada has a highly mechanised form of commercial arable farming.

(b) (i) Descriptions might include:
- improved irrigation.
- increased farm sizes and larger fields.
- increased use of fertilizer.
- increased mechanisation.
- 'green revolution' type changes eg development of hybrid seeds.
- use of appropriate technology.
- increasing export of farming produce.

(ii) The impact of these changes might include:
- greater amount of food has reduced malnutrition and starvation.
- surplus crop may be sold, improving quality of life.
- increased mechanisation may lead to reduction in farm labour.
- migration of farm workers to urban areas and impact on demography of rural areas.
- consolidation of farms may also lead to larger fields, increased mechanisation and drift to cities.
- improved infrastructure including increased electrification and better roads improving access to markets.
- larger, more effective irrigation schemes and drainage systems.
- increased use of insecticides, pesticides and fertilisers may impact on the environment and humans.

Question 4 – Industrial Geography

(i) The following points might be included in a description/explanation of the major characteristics of a "new" industrial landscape.
- Lower, smaller, modern buildings – mostly single storey and often with large windows to allow in plenty of light.
- Buildings are well planned/spaced out with trees and grassy areas and even ornamental lakes/ponds included in the layout to provide a more attractive working environment and create a favourable image to prospective investors/clients.
- These areas are usually located on purpose-built industrial estates or Science/Business Parks commonly on Greenfield sites on the edge of towns/cities where land is relatively cheaper and there is room for car parking and for future expansion.
- They are close to major roads such as dual carriageways or motorways for ease of transport of the finished products to markets/ports, for bringing in raw materials/components/sub assemblies and for the convenience of to-day's more mobile, car-owning workforce.

- There is an absence of slag heaps/coal bings, factory chimneys, railway sidings etc usually associated with older, 'smokestack' industrial areas.
- There is a tendency for similar sorts of industries/firms in similar looking buildings to be located on the same site to benefit from an exchange of ideas and information. Many of these businesses are connected with information, high technology and electronics industries and will have direct links with universities (often situated close by) for research and development purposes and to remain successful and competitive.

(ii) Answers will, of course, depend on the industrial concentration chosen.
National government measures may include:
- incentives such as capital allowances, retraining grants, provision of purpose-built premises/advance factories, rent-free periods, tax relief on new machinery, reduced interest rates…
- specific financial assistance to old industrial areas such as coalfields in UK.
- the creation of Enterprise Zones, Development Areas, Urban Development Corporations. Regional Selective Assistance (Scotland/Wales). Selective Finance for Investment (SFFI – England).
- setting up Manpower Creation Schemes (MCS) and Youth Training Schemes (YTS).
- decentralising/relocating government offices/departments (eg DVLA to Swansea or National Savings to Glasgow).
- Improving communications/accessibility/existing infrastructure.

EU assistance could include mention of such agencies as ERDF (European Regional Development Fund), EIB (European Investment Bank), ESF (European Social Fund), etc and their associated benefits and/or more general references to EU funding/regeneration projects.

SECTION B

Question 5 – Atmosphere

(a) Description and Explanation might include:
- currents follow loops or gyres – clockwise in the North Atlantic. In the Northern Hemisphere the clockwise loop or gyre is formed with warm water from the Gulf of Mexico (Gulf Stream/North Atlantic Drift) travelling northwards and colder water moving southwards eg the Canaries Current.
- currents from the Poles to the Equator are cold currents whilst those from the Equator to the Poles are warm currents. Cold water moves southwards from Polar latitudes – the Labrador Current. This movement of warm and cold water thus helps to maintain the energy balance.
- ocean currents are greatly influenced by the prevailing winds, with energy being transferred by friction to the ocean currents and then affected by the Coriolis effect, and the configuration of land masses which deflect the ocean currents. Due to differential heating, density differences occur in water masses, resulting in chilled polar water sinking, spreading towards the Equator and displacing upwards the less dense warmer water.

(b) Candidates should be able to name and explain the mechanism of each of the three cells – Hadley, Ferrel and Polar – and should describe their role in the redistribution of energy.

Eg warm air rises at the Equator, travels in the upper atmosphere to c.30°N and S, cools and sinks. Some of this air returns as surface NE or SE trade winds to the Equator to form the Hadley Cell.

The remainder of the air travels north over the surface as Westerlies to converge at about 60°N and S with cold air sinking at the Poles and flowing outwards. This convergence causes the air to rise – some of this air flows in the upper atmosphere to the Poles where it sinks forming the Polar Cell. Candidates may note the Easterlies from the High Pressure area at the Pole.

The remainder of this air in the upper atmosphere travels south and sinks at 30°N and S to form the Ferrel Cell. Credit should be awarded to candidates who recognise that the eastward passage of depressions and associated jet streams deforms any Ferrel Cell out of recognition.

It is in this way that warm air from the Equator is distributed to higher (and cooler) latitudes and cold air from the Poles distributed to lower (and warmer) latitudes.

Question 6 – Lithosphere

(a) Evidence which suggests that Area A on Reference Map Q6 is a Carboniferous Limestone landscape could include:
- extensive areas of bare rock – scars, crags, limestone pavements (eg around and to the west of Malham).
- numerous mentions of underground features such as caves and pot holes (eg those in squares 8570 and 8569) formed due to limestone's susceptibility to solution by acidic rainwater.
- absence of surface drainage over large areas suggesting permeable rock.
- disappearing streams like the one from Malham Tarn which sinks at 894657 and appears to resurface further south below the high cliff of Malham Cove.
- gorges such as Gordale Scar (9164) which, it has been suggested, may have been formed by the massive collapse of cavern roof systems.

(b) Probably the most obvious (sensible!) Carboniferous Limestone feature to choose would be a limestone pavement although some candidates may focus on limestone caves and their associated underground landforms such as stalactites, stalagmites and rock pillars.

In **explaining** the formation of a **limestone pavement**, for example, candidates could refer to such points as:
- the part played by glacial erosion (abrasion) in scraping away any overlying soil cover and thus exposing the horizontally-bedded, rectangular blocks of limestone.
- joints formed in the limestone as it dried out and pressure was released.
- these joints/lines of weakness are more prone to chemical weathering than the surrounding limestone. The limestone is dissolved over time by rainwater (weak carbonic acid) leaving deep gaps (grykes) and intervening blocks (clints).
- continued weathering (both physical and chemical) will further deepen and widen the grykes.

SECTION C

Question 7 – Population

(a) Description could include:
- decreasing birth rate – narrow base and low proportion in youngest age groups.
- decreasing death rate – pyramid narrows naturally from 60's age group and has an inverted pyramid shape lower down.
- increasing life expectancy – many more in the elderly section of the population and a significant number over 100, especially females.
- economically active population decreasing – fewer in the 15 to 65 age groups by 2050.

Explanation could include:
- decreasing birth rate – people wanting smaller (cheaper) families, women following careers, easier family planning/contraception/abortion.
- Decreasing death rate and increasing life expectancy – improved health care, sanitation, housing, food supply, pensions, care for the elderly.

(b) Possible consequences could include:
- the decline in the birth rate may lead to less demand for services/industries needed for the smaller child population with the ensuing problems caused when these are closed or scaled down (eg schools or nurseries).
- the 'greying' of the population may lead to a need for more geriatric care with increased strain/costs on health centres/local authorities/central government.
- a decrease in the economically active population may lead to key jobs not being filled and higher taxes from the working population to pay for their elderly dependents. The qualifying age for pensions may have to be raised and the government may have to reduce their state pension whilst encouraging more private pension plans/health care etc.
- the government may need to encourage the inward migration of key workers that may lead to cultural/language/religious difficulties.
- the government may also need to encourage a higher birth rate.

Question 8 – Urban Geography

(i) For Glasgow candidates may refer to:
- loss of custom for shops in the CBD due to competition from out-of-town shopping centres like Braehead with their large car parking areas.
- consequent closure of shops, especially at the less profitable edges of the traditional CBD due to reduced pedestrian flow, eg High Street end of Argyle Street, giving empty shop units and 'run-down' appearance.
- revitalisation of shopping centres in central CBD – eg building of Buchanan Galleries and renovation of St Enoch Centre in order to compete/keep up.
- shops in CBD may be less overcrowded at peak times, eg Christmas, giving improved shopping experience at these times.
- focus on designer label/high-end shopping taking advantage of CBD status, eg Princes Square, Italian Centre.

(ii) For Glasgow, candidates may refer to:
- pedestrianisation and landscaping of CBD roads eg Buchanan Street, Argyle Street etc to reduce traffic flow in and around the CBD – to increase pedestrian safety and improve air quality and environment. Upgrading of CBD open space, eg George Square.
- diversification of city employment – much greater emphasis on tourist industry (significance of city-break holidays) leading to increased bed accommodation in new CBD hotels (Hilton, Radisson). Hotels can also tap into lucrative conference market given Glasgow's improved image as a tourist and cultural centre.
- alteration of CBD road network – one-way streets (around George Square), bus lanes to discourage use of private transport and encourage use of public transport. Also achieved by increased metering and increased parking charges in and around CBD.
- renovation and redevelopment of many CBD sites to provide modern hi-tech office space (Lloyd's TSB, Direct Line etc) and residential apartments (Fusion Development, Robertson Street).
- building of M8, M77 and M74 extension all designed to keep through traffic off CBD roads.
- younger, more affluent population continues to be attracted to central city area by long-standing concentration of up-market pubs, clubs, cinemas etc (Cineworld in Renfrew Street).
- contraction of number of public transport termini within CBD (2 major railway stations instead of 4) but upgrading of remaining termini, (Buchanan Street bus station, Central Station).

GEOGRAPHY HIGHER ENVIRONMENTAL INTERACTIONS 2009

Question 1 – Rural Land Resources

(a) For a corrie points could include:
- snow accumulates in a (north-facing) hollow on mountainside
- successive layers of snow compress first snowfalls into ice/neve
- ice moves downhill under gravity
- freeze-thaw weathering loosens rock above glacier
- plucking steepens back wall of the corrie
- maximum erosion takes place where weight of ice is greatest
- boulders and stones embedded in ice grind away at the bottom of corrie
- abrasion carving out hollow/armchair shaped depression (over deepening)
- rate of erosion decreases at edge of corrie leaving a rock lip.

Minimum of 3 features needed for full credit.
For an answer to achieve full marks well annotated diagrams must be used.

(b) Responses will vary according to the area chosen but opportunities may include:

tourism, recreation, nature conservation, hill farming, forestry, HEP generation potential, water supply, quarrying. For forestry, candidates could describe the use of plantations for providing wood for furniture, building materials, Christmas trees (economic) and also forest walks, nature trails, picnic sites, mountain bike and orienteering courses (social) with resulting employment (reducing rural depopulation) and profits/taxes for businesses/government.

(c) (i) Conflicts may include:
- traffic congestion especially on narrow rural roads and in car parks - particularly during peak holiday periods
- large volumes of visitor traffic increase air and noise pollution and can spoil the attraction of many villages
- increase in holiday home ownership has pushed up prices forcing many local people to move away
- erosion of footpaths and resulting visual pollution due to high visitor numbers
- some visitors may cause problems for farmers and landowners (eg damage to property such as stone dykes, animal disturbance)
- development of unsightly visitor/leisure complexes/caravan sites etc.

(ii) Solutions might include a variety of environmental conflicts depending on the area chosen but for tourism/traffic it could be:
- traffic restrictions in more favoured areas/at specific peak times, eg one-way streets, bypasses or complete closures
- encourage the use of public transport eg park and ride, minibuses and the use of alternative transport eg cycle paths and bridle ways
- separating tourist and local traffic, the use of permits (for access or parking) in some areas.

Mention of the success/failure of the solution is required for full credit.

Question 2 – Rural Land Degradation

(a) Answers should be able to pick out such points as:
- the considerable fluctuations in rainfall over the period shown
- the preponderance of wetter than average years – certainly between the end of the 19th century and 1970
- the marked concentration of wet years between 1950 and 1970 (only 2 drier than average in these two decades – 1960 and 1969)
- the very obvious pattern of drier years throughout the 1970's and 1980's
- the marked differences in more recent years.

(b) (i) For **Africa north of the Equator** human activities/inappropriate farming techniques contributing to land degradation could include:
- overgrazing; over cropping; deforestation; monoculture; reduced fallow periods; the growing of cash crops; increased population pressure leading to more and more marginal land (more vulnerable to erosion) being cultivated.

For the **Amazon Basin**:
- deforestation – eg for logging/ranching/mineral extraction/road building HEP projects/ resettlement schemes/charcoal burning
- the impact of ranching – forest cleared, used for a few years until grass fails – move and clear a new area of forest and so perpetuate the whole process
- population pressure and shortage of available land arising from these increased demands on the rainforest have resulted in some shifting cultivators returning to tribal lands prematurely and therefore encouraged soil erosion.

(ii) For **Africa north of the Equator** *human* consequences of land degradation could include:
- crop failures/death of livestock – reduced food supply – malnutrition – famine – increased infant mortality rates and death rates
- encourages the drift from rural areas to already overcrowded urban areas – growth of shanty towns
- the collapse of traditional nomadic ways of life

Whilst *environmental* consequences might be:
- soil structure breaks up due to over cropping and monoculture
- wind erosion can remove dried out soil
- deforestation means that soil is exposed and more prone to erosion
- torrential rain often leads to widespread gullying which is impossible to rectify – further loss of precious farmland
- level of water table reduced
- intensified drought due to albedo effect.

For the **Amazon Basin** *human* consequences could include:
- destruction of the way of life of the indigenous population
- clashes between tribal groups and 'outsiders'/developers
- creation of reservations for indigenous people
- impact of "western diseases" on tribal people
- encourages rural-urban migration

Whilst *environmental* consequences might be:
- adverse effect on the rainforests closed nutrient cycle
- leaching of minerals/removal of topsoil/increase in laterisation with loss of protective natural vegetation cover
- increased run-off and flooding; silting up of rivers
- loss of wildlife habitats/biodiversity – some species threatened with extinction
- loss of potentially useful medicinal drugs
- impact on global climate (Greenhouse Effect).

(c) (i) Soil conservation measures employed in **North America** could include:
- contour ploughing
- crop rotation/diversification

- trash farming/stubble mulching
- planting of shelter belts/windbreaks
- strip cultivation
- increased use of irrigation.

(ii) eg in relation to **shelter belts** – trees planted at right angles to the direction of the prevailing wind have managed to reduce wind speeds and provide an effective barrier to protect the soil. The taller and more complete the tree cover, the more effective the shelter. They also improve water retention and help bind the soil together. The negative consequences of this might also be noted eg the area of land occupied by trees/hedges and competition for water and nutrients.

For full credit candidates should mention a minimum of 4 strategies. Some comment on the effectiveness of each measure is also required for full credit.

Question 3 – River Basin Management

(a) Candidates may mention a range of reasons to explain the need for water management including:
- low annual precipitation
- flood control
- regulating flow and storage of water
- power supply for expanding cities and industry
- water for industry
- water for irrigation as food demands increase
- drinking water for increasing population.

(b) Physical factors might include:
- solid foundations for dams
- consideration of earthquake and underground movements
- narrow cross-section to reduce dam length
- large, deep valley to flood behind dam
- impermeable rock beneath reservoir
- sufficient water supply from catchment area
- low evaporation rates, due to small surface area of reservoir
- impact on hydrological cycle.

Human factors might include:
- cost of construction
- proximity of urban areas for water and electricity
- proximity of agricultural areas for irrigation
- cost of displacing people
- cost of compensating farmers and home owners
- impact on communications.

(c) (i) Problems might include:
- difficulties between states which are represented by different political parties
- sharing allocation of water rights
- changing needs of different states including increasing populations and increasing irrigation
- increased pollution and salinity downstream affecting water quality
- shared costs of purification and desalination plants
- impact of dam construction on consumers downstream
- relationship between neighbouring countries.

(ii) These problems could be overcome by having political agreements eg the Colorado River Compact which divided up water allocations based on historical rainfall patterns. International agreements may be needed where different countries are involved.

(d) Social benefits include:
- improved water supply for drinking
- irrigation providing an increased food supply
- less water borne disease
- population increases sustainable
- greater availability of electricity
- opportunities for tourism and recreation
- increased fresh water improves health and sanitation.

Economic benefits include:
- improved navigation and roads across dams
- HEP and water for industry
- irrigation allowing double cropping and commercial farming
- income from tourism and recreation.

Environmental benefits include:
- flood control
- reliable seasonal water supply
- dramatic scenery around dams and reservoirs
- introduction of new wildlife habitats.

Candidates must refer to all 3 sections for full marks

Question 4 – Urban Change and Management

(a) Answers will depend upon the EMDC chosen, but for Spain answers might suggest:
- concentrations on and around the coast of Spain, both on the North coast and along the East and South coastal areas
- coastal cities would include ferry terminals (Santander, Palma), historical trading ports (Barcelona) and holiday areas accessed by airports (Malaga, Alicante)
- along rivers for communication, trade, raw materials (Sevilla, Zaragoza)
- major cities on Spanish islands (Palma, Mallorca and Las Palmas, Gran Canaria)

(b) (i) Social, Economic and Environmental problems should be related to the candidate's chosen city. Answers would be enhanced by convincing relevant details on the chosen city such as **named** shanty areas or specific projects to tackle problems.

Problems might include:
- continued growth of these shanty towns (favelas, bustees etc) in and around the city
- shanty areas are characterised by poor quality home-made dwellings, overcrowding, inadequate water and power supplies, poor sanitation, disease and general lack of amenities like services, schools and hospitals
- they are often sited on unstable hillsides, marshy areas or other areas avoided by other building
- unemployment or underemployment and poor wages for the few jobs available
- 'grey' or 'black market' economies with problems of drugs and high crime rate
- chronic traffic congestion and high pollution levels from nearby industries
- sites are illegally settled and may be bulldozed and removed by city authorities at any time.

(ii) Ways to tackle problems might include:
- self-help schemes (eg São Paulo) where city authorities provide basic housing made of breeze block and roof tiles. Local residents supply the labour for 'finishing off' and digging ditches for water supply and sanitation
- basic amenities such as power, clean water, roads and community facilities may be provided
- groups of residents within shanty town areas may form community groups to share trade skills to improve existing facilities within the larger shanty town
- city authorities may build high-rise apartment blocks in suburbs to provide high-density housing to replace the extremely high-density living in shanty areas

Some qualitative statements on the success or otherwise of these schemes is required to attain full marks. For example, "the advantages of self-help schemes are that costs are kept to a minimum to maximise the number of 'basic shell' houses that can be built. Working together can establish community spirit with shared common purpose."

(c) (i) Maximum of 14 in (b) (i) and (ii) – if named city.

Traffic congestion in an EMDC city. For Aberdeen candidates might suggest:
- increased commuting from dormitory towns and villages to N, W and S of Aberdeen as people seek quieter living conditions focuses rush hour traffic on major traffic junctions eg Haudagain roundabout
- Aberdeen has major industrial areas at Altens and Dyce, leading to large commuter flows outwith the city centre
- major roads have to converge to cross the rivers Don and Dee, leading to bottlenecks at bridges over the rivers
- around 15,000 journeys per day in Aberdeen are generated by through traffic, clogging up city streets unnecessarily
- more stringent traffic regulations in and around the CBD and shortage of car parking facilities, leading to unnecessary traffic flow as spaces are sought
- growing car ownership, related to high disposable incomes and increased number of 2 (or more) car families. > 60% of employed people in Aberdeen travel to work by car
- increased use of private transport to do 'school run' during rush hours, due to safety issues
- increase in number and size of lorries and buses which often find it difficult to manoeuvre in outdated road network, delaying other traffic
- increase in need for road maintenance due to increased traffic flow and weight of modern lorries
- shutting off of side roads formerly used as 'rat runs' focuses all traffic on to main arterial roads.

(ii) Candidates may suggest protests and land-use conflicts that would be due to:
- breaching of green belt land
- roads such as the AWPR use up large tracts of land, often good quality farmland or recreational land, leading to protests by groups such as the Aberdeen Greenbelt Alliance or Friends of the Earth who want to preserve green belt areas
- removal of sensitive woodlands and meadows in the Dee Valley may harm endangered species in the areas, such as the Red Squirrel
- once roads such as the AWPR have been built, they act as a focal point for developers wanting to build houses and/or industrial areas and/or out-of-town retail parks with vast buildings and huge car parks taking up large areas of land, these all benefit from the improved access brought by the new road, these also in turn lead to further breaching of the green belt land, including potential loss of golf courses, country parks etc
- property blighting, where any properties which are within sight or sound of the road may lose much of their value
- spiralling costs of projects like the AWPR lead protesters to argue that the huge amounts of money involved could be used to improve existing infrastructure and modernise public transport
- compulsory purchase orders are placed on properties on the proposed route and people are forced out of their houses.
- farms are often split by such roads, causing access problems for farmers and their livestock
- encroachment of housing may increase crime and vandalism for farmers and their property.

Question 5 – European Regional Inequalities

(a) Candidates should note the higher levels of development in pre-2000 states. This may be due to locations near the economic Core, the length of membership and subsequent financial benefits, the quality of infrastructure, the proximity of markets, the ease of trade and the sourcing of resources. The relative prosperity of the pre-2000 states also leads to a better quality of life. Candidates will also gain credit from noting anomalies and where possible outlining reasons for these anomalies eg Portugal and Slovenia.

(b) (i) Credit should be awarded for candidates noting that Objective 1 status is awarded to Europe's peripheral areas. A maximum of 6 marks should be awarded for identifying areas or countries eg Ireland, SW England, Wales, Portugal and Spain, S Italy, Greece, Sweden, Finland, the eastern part of Germany and the new states of the Eastern bloc.

(ii) Countries would benefit from Objective 1 status through the following:
- support for infrastructure improvements
- support for employment training and education
- support for production/manufacturing sectors
- environmental protection
- improving access to the peripheral areas
- improving IT, literacy and numeracy.

(c) (i) Answers will be dependent upon the country chosen but must be authentic for the candidate to score full marks.

Physical factors could be related to (for example):
- difficult terrain/relief (mountain ranges: Northern Spain, Central Italy, Highlands of Scotland)
- physical isolation (Highlands of Scotland/Wales, areas in Central Spain, much of Southern Italy)
- climatic problems such as drought (Southern Italy, Central, Southern and Eastern Spain, Greece)
- prolonged winters (Northern Sweden, much of Finland)
- natural disasters (earthquakes; Greece, Italy; volcanic eruptions; Southern Italy).

Human factors could include (for example):
- differences in employment opportunities
- decline in the range and scope of opportunities in the rural-based economy
- decline in range of relevant skills in a declining industrial area
- perceptions of inward investors
- problems of land ownership and tenure
- political/terrorist factors (ETA)
- overdependence on seasonal employment in (for example) the tourist industry
- variations in investment infrastructure
- distance from main markets/Euro-core.

(ii) Once again answers will depend upon the country chosen but national government strategies to tackle regional inequalities might include:
- regional development status, Enterprise Zone status, capital allowances, training grants, assistance with labour costs
- specific assistance to former coal mining/iron and steel areas
- intervention of national government resulting in the location of major government employers in disadvantaged areas DVLA (Swansea), MOD (Glasgow), SNH (Inverness)
- intervention by national government to encourage inward investment – particularly in newer industries – electronics/call centres.

Question 6 – Development and Health

(a) Essentially single indicators are too broad/generalised:
- they are averages which disguise or distort wide internal variations eg a few immensely wealthy families but the majority of the population may be living at subsistence level
- combining indication on health, education and the economy give a more balanced view of development
- some regions/areas of a country may be much better off than others – 'north-south' or 'urban-rural' contrasts
- GNP figures are in some cases inflated by oil revenues (showing a big gap between these and other indicators that have yet to 'catch up')
- subsistence agriculture and 'barter economies' are not included in wealth indicators
- certain indicators are perhaps irrelevant to the real quality of life in many poorer ELDCs eg TVs per household when there is no electricity supply.

(b) Differences in the levels of development between ELDCs (Economically Less Developed Countries) may be due to:
- mineral reserves eg Saudi Arabia and similarly positioned Middle East countries with vast reserves of oil. They also have stable (if despotic) government regimes/monarchies that leads to the generation of huge wealth. This wealth can 'trickle down' to a wide sector of the population. Other countries may have no reserves of minerals in demand by the EMDCs eg Burkina Faso
- political instability eg many have unstable regimes or are suffering from border wars and/or civil wars eg Zimbabwe, Sudan and Indonesia
- colonial links eg some Caribbean countries receive support from European countries because of their former colonial ties
- strategic locations eg South Korea and many Central American countries receive additional support from leasing land for military bases
- encouragement of entrepreneurial skills and the ability to attract in major world companies eg by offering an educated, resourceful and relatively cheap work force (South Korea) and/or incentives eg 10 years rent free factory sites in Vietnam
- natural disasters eg Bangladesh (cyclones and floods), Niger (recurring drought and associated famines) will limit progress as development money is spent on repairing infrastructure and humanitarian aid.

(c) (i) For **Malaria** – Environmental factors:
- suitable breeding habitat for the female Anopheles mosquito – areas of stagnant water to lay eggs in
- hot and wet climates such as those experienced in the Tropical Rainforests or Monsoon areas of the world
- temperatures between 15°C and 40°C
- areas of shade in which the mosquito can digest blood.

Human factors:
- nearby settlements to provide a 'blood reservoir'
- encouraged by bad sanitation and poor irrigation or drainage that leaves standing water uncovered eg tank wells, irrigation channels, water barrels, padi fields.

(ii) Strategies used to combat the spread of Malaria can include:

Trying to **eradicate** the mosquito:
- insecticides eg DDT and now Malathion
- mustard seeds thrown on the water that become wet and sticky so dragging the mosquito larvae under, drowning them
- egg-white sprayed on the water creates a film which suffocates the larvae by clogging up their breathing tubes
- bti bacteria grown in coconuts – the fermented coconuts are broken open after a few days and thrown into the mosquito infested ponds. The larvae eat the bacteria and have their stomach linings destroyed
- larvae eating fish introduced to ponds
- draining swamps, planting eucalyptus trees that soak up excess moisture, covering standing water
- genetic engineering eg of sterile males

Treating those suffering from malaria:
- drugs like chloroquin, larium and malarone
- quinghaosu extracted from the artemesian plant – a traditional Chinese cure
- continued search for a vaccine – not available as yet
- education programmes in –
  - the use of insect repellents eg Autan
  - covering the skin at dusk when the mosquitoes are most active
  - sleeping under an insecticide treated mosquito net
  - mesh coverings over windows/door openings
- WHO 'Roll back malaria' campaign
- the Bill and Melinda Gates Foundation

(iii) Benefits to ELDCs of controlling disease may include:
- funds can be diverted elsewhere in the Health sector or transferred to other budgets that help development
- national debt can be reduced
- the workforce will be fitter (eg farmers better able to produce food), thus also helping to raise health levels
- productivity will increase as the workforce takes less sick leave/life expectancy increases
- the area will become more attractive to tourists, foreign currency income can be generated and this will also assist in developing tourism related services/industries
- a possible reduction in birth rates as a result of a fall in infant mortality rates.

(d) Examples of Primary Health Care (PHC) strategies may include:
- use of barefoot doctors – trusted local people who can carry out treatment for more common illnesses – sometimes using cheaper traditional remedies
- use of ORT (Oral Rehydration Therapy) to tackle dehydration – especially amongst babies. This is an easy, cheap and effective remedy for diarrhoea/dehydration
- provision of vaccination programmes against diseases such as polio, measles, cholera. Candidates may also refer to PHC as based on generally preventative medicine rather than (more expensive) curative medicine
- the development of health education schemes in schools, community plays/songs concerning AIDS, with groups of expectant mothers or women in relation to diet and hygiene. Oral education being much more effective in illiterate societies
- sometimes these initiatives are backed by the building of small local health centres staffed by doctors (like GPs)
- PHC can also involve the building of small scale clean water supplies and Blair toilets/pit latrines – often with community participation
- The use of local labour and building materials is often cheaper, it also provides training/transferable skills for the participants and gains faster acceptance/usage in the local and wider community.

Explanation of why PHC is more effective in ELDCs is required to gain full marks.

GEOGRAPHY HIGHER
PHYSICAL AND HUMAN ENVIRONMENTS
2010

SECTION A

Question 1 – Atmosphere

(a) Human factors
- Carbon Dioxide: from burning fossil fuels – road transport, power stations, heating systems, cement production and from deforestation (particularly in the rainforests) and peat bog reclamation/development (particularly in Ireland and Scotland for wind farms).
- CFC/PFCs: from aerosols, air-conditioning systems, refrigerators, polystyrene packaging etc.
- Methane: from rice paddies, animal dung and belching cows.
- Nitrous oxides: from vehicle exhausts and power stations.
- Sulphate aerosol particles and aircraft contrails: global 'dimming' – increase in cloud formation increases reflection/absorption in the atmosphere and therefore cooling.

NB There were 6 man-made greenhouse gases included in the Kyoto protocol (Carbon Dioxide, Methane, Nitrous Oxide, Hydroflurocarbons, PFCs and Sulphur Hexafluoride). Many more powerful than CO_2.

(b) Answers may include:
Melting of the ice sheets/glaciers
- A rise in sea level with subsequent migration as islands and coastal areas are submerged. Loss of plant and animal habitats in these areas eg impact on polar bears which could lead to a loss of tourism/more problems in settlements as the bears scavenge instead of hunting on the ice.
- New transportation routes across the Arctic Ocean ie the North West Passage with resulting benefits to trade/previously ice bound coastal settlements.
- Extension of mineral exploitation into the Arctic with positive and negative consequences.

Changing rainfall/temperature patterns
- Higher or lower rainfall/temperature and maybe more extreme weather depending on where you are with resulting increasing/decreasing crop yields, more floods/drought/hurricanes/tornadoes etc.
- Extension or retreat of vegetation (and associated wildlife) by altitude as well as latitude – growing vines/sunflowers in Scotland, spread of malaria, the loss of the Cairngorm Arctic habitat etc.
- Change in ocean currents (El Nino/La Nina).
- Change to the Atlantic Conveyor – disruption of the thermohaline circulation.

Question 2 – Lithosphere

A sequence of diagrams, fully annotated, could score full marks in either part (a) or part (b).

(a) Conditions and processes which encourage the formation of scree slopes.
- Steep and bare rock faces with lines of weakness/well-jointed carboniferous limestone.
- Cold climate where temperatures often fall below freezing point at night.
- The two factors above allow physical weathering to take place in the form of freeze-thaw action/frost shattering, where water collects in the rock fractures, freezes and expands by about 9% exerting great pressure on even the hardest rock.
- Repeated freeze-thaw action splits the rock into large sharp fragments which break off and are moved downhill by gravity to accumulate at the base of steep slopes as a scree or talus slope as large heaps of rock debris.

Full marks can only be achieved if one or more annotated diagram is included.

(b) Processes involved in formation of a corrie.
Corries
- Snow accumulates in mountain hollows when more snow falls in winter than melts in the summer.
- North/North-east facing slopes are more shaded so snow lies longer.
- Accumulated snow compresses into neve and eventually ice.
- Plucking, when ice freezes on to bedrock, fractures it and incorporates it into the glacial ice.
- Abrasion, when the angular rock within the glacial ice grinds away the valley sides and floor, over-deepening the hollow along with rotational movement of glacier.
- Glacier moves downhill due to gravity.
- Rotational movement not so powerful at corrie edge, allowing rock lip to form which traps water as ice melts, leaving a lochan.

Question 3 – Population Geography

(a) Answers could include points such as:

Description	Explanation
The largest number of migrants come from Poland (124,000)	Due to the expansion of the EU in 2005 and freedom of movement for workers
The second highest source was India (about 100,000)	Possibly due to Commonwealth links or people who already have family in the UK's large Indian sector
A large proportion (48,000) came from Australia	Due to the lack of a language barrier and the increasing trend towards young people travelling for work experience

Reference can also be made to relevant push and pull factors.

(b) Answer will depend on migration studied.
A developed answer will refer to both advantages and disadvantages for the country.
From Mexico into California, answers may include:

Mexico

Advantages	Disadvantages
The pressure on resources and jobs was lessened.	The active population left, creating a burden on the economy.
The birth rate was also lowered as most migrants were of childbearing age.	Those most educated left creating a 'brain-drain'.
Money was often sent back to the families left behind, which helped to stimulate the economy – it is Mexico's biggest source of foreign income ($6 bn per year).	Families are divided as males leave.
When migrants return, they can bring back new skills, which can be used in the donor country.	Death rate increases as an elderly population is left.
	In the long term this creates dependency upon money sent back to home villages.

California

Advantages	Disadvantages
• The short-term labour gap was filled – migrants filled jobs Americans did not want. • Mexican culture has enriched the border states with language, food and music. • Increased population leading to increased taxation levels. • Labour costs reduced – agricultural sector benefits from this.	• Migrant workers feel discriminated against and socio-economic problems have ensued. • When recession hit in the 1980s unemployment rose and racial tension was exacerbated. • Ghettos developed in the poorest districts. • TB has increased along the border. • Illegal migration costs the USA millions of dollars for border patrols and holding centres.

Question 4 – Urban Geography

(a) The following characteristics may be noted:
  • Densely packed and irregular street pattern.
  • Transport centres eg bus station and railway station.
  • Bridging points across River Ouse.
  • Historical buildings eg The Minster, Castle.
  • Important buildings eg information centre, churches and Town Hall.
  • Evidence of inner ring road.

(b) The advantages of the residential location and environment may include:
Area B (suburban housing area – Rawcliffe).
  • Access to A19 for commuting to CBD.
  • Near park and ride for commuting, and National cycle route.
  • Modern design of cul-de-sacs and crescents for privacy and preventing through traffic, and roundabouts at access points.
  • Services including a church for local use.
  • Near industrial estate GR593553 for employment.
  • Tourist facility to east ie Nature Reserve, and caravan site.
  • Attractive environment ie small lake, on edge of town near farmland.
  • Clifton Moor Retail Park 591558.

Area C (commuter village – Copmanthorpe).
  • 6km from centre of York for shopping, work and entertainment.
  • Nearby sliproad onto A64, ideal for commuters.
  • Small, quiet village with a few services eg post office, public house, church.
  • Leisure facility to north ie golf course.
  • Environmentally attractive with Ebor Way going through the village and Askham Bogs Nature Reserve to the north.
  • Surrounded on three sides by farmland.

(c) Candidates should be able to demonstrate an understanding of the issues which arise when 'urban' land uses invade a previously 'rural' area.
Land uses which would conflict with further expansion would include
  • National walking and cycle trail.
  • New shopping centre – expansion may be restricted.
  • Leisure facilities eg racecourse, golf course.
  • Various farms eg White House farm.
  • A64 bypass.

  • Accommodation including Manor Hotel, caravan and camping site GR600476.
Other land uses include forestry, small villages, college, university, electricity transmission lines.

SECTION B
Question 5 – Hydrosphere

Full marks can only be achieved if one or more annotated diagram is included.

(a) The explanation should include 8 points, all of which could be included in a well annotated diagram.
Points could include:
  • Development of pools and riffles (differences in speed and depth).
  • Erosion on the outside (concave bank) of bends due to faster flow.
  • Helicoidal flow removing material.
  • Deposition on the inside (convex bank) due to slower flow.
  • Formation of point bars.
  • Migration of meanders downstream.

(b) The physical factors may have included:
  • The close proximity of a tidal limit to York, GR594445.
  • A low lying floodplain around or below 10 metres in height.
  • Tributaries joining the River Ouse eg River Foss (605510).

The human factors may have included:
  • Home building on a flood plain.
  • River straightening.
  • Man-made strengthening and narrowing of river banks.
  • Facilities deliberately located on floodplain eg racecourse, caravan and camping site.
  • Land use changes in catchment.
  • Increased run-off from the 'urbanised' environment.

Question 6 – Biosphere

Students may answer this question using the headings on the graph or the more usual progression from strandline to climax vegetation.

• **Plant cover increases** – the amount of sand showing through the dune decreases as more of the ground is covered by vegetation. Little cover in the pioneer stage, more in the building stage and complete cover in the climax stage unless disturbed by animals/humans/storms etc.
• **Soil moisture increases** – rain/fresh water is trapped with added humus/plant cover and longer rooted plants drawing water up from the water table. Xerophytic plants found in the drier strandline (sea sandwort, sea rocket, saltwort) and embryo dune (sea or sand couch, lyme grass, frosted orache). On the yellow dune Marram grass has long rhizomes to spread through the sand. Dune slacks at or near the water table have hydrophytic species like reeds, rushes and flag iris.
• **Organic matter content increases** – decaying pioneer species adding humus to the sand. In the fore-dune more plants stabilise the sand adding humus (sea bindweed, sea holly, sand sedge, and marram grass) changing the sand to a sandy loam and from the sandy colour of this and the yellow dune to the grey dune.
• **PH decreases** – shells (CaCO3) producing alkaline conditions on shore, more neutral pH by the climax stage as plants decay and add acid to the soil. The grey dune plants include sand sedge, sand fescue, bird's foot trefoil, heather, sea buckthorn and grey lichens. In the climax stage a range of plants from heathers to birch, pine or oak woodland can grow depending on the final pH value of the soil.
• **Salinity decreases** – Increased distance from the sea and salt water/tides/spray increases the amount and variety of plant species that can cope with the conditions.

SECTION B
Question 7 – Rural Geography

(a) Candidates may answer this question in one or two parts. Answers could include the following:

(i) Mechanisation increases the efficiency on a farm enabling the farmer to plough, sow, spray, etc more quickly, covering larger areas. It also speeds up harvesting and results in the product being delivered to markets fresher and at a higher premium (eg Bird's Eye peas). It also allows for a smaller work force and therefore lower wage bills for the agribusiness usually involved in these farms. It allows for the use of satellite technology/ computers to control the application of fertilisers to particular areas of fields to improve yields (yet decrease the cost and waste) as only the required amounts are delivered to each segment according to the soil quality there.

(ii) **People in the environment** – this leads to depopulation and derelict buildings, deserted rural villages (eg on the Great Plains).
Farm sizes – increasing 'agribusiness' type farming with amalgamated farms, larger fields, fewer hedgerows or boundaries to allow for machinery and increase yields. More or larger buildings for storage of machinery.
Pollution – air and water pollution from machinery itself (exhaust fumes/noise/accidents with diesel) and run-off from over application of fertilisers.

(b) For EU Policies eg Single Farm Payment:
This payment has replaced the existing support schemes to farmers like the arable area payment scheme and includes other entitlements like the set-aside entitlement.
The total payable to the farmer is calculated using the historical payments made to the farm from 2000 to 2002. The farmer gets a single payment based on these calculations. To continue to receive this payment the farmer must keep the land in 'good agricultural and environmental condition' – this includes an obligation to keep land in set-aside. Penalties will be made if these conditions are not met – government inspectors will visit to check!

For Genetically Modified food/genetic engineering:
Scientists manipulate the genes of plant cells by adding/deleting DNA. The first commercial genetically engineered food was the Flavr Savr tomato – by adding a fish gene it had a longer 'shelf-life'. Since then the developments have concentrated on four main crops – oilseed rape, cotton, maize and soya bean. These crops have been made herbicide resistant (they can tolerate the herbicides that will kill the weeds) and/or pest resistant (they produce a toxin that kills the pest that would normally eat them). This technology has also been used to improve the taste, nutrition or drought-resistance of the crop. Ethical concerns, health risks, environmental concerns eg cross-pollination to produce 'superweeds' have led to these crops being limited in area (although that is still estimated at 1 million square kilometres).

Question 8 – Industrial Geography

(i) Answers will depend on the chosen area. For the industry in the question.
- Increased competition from overseas markets, particularly NICs.
- Increased competition leading to falling prices and profits.
- Falling customer demand for the product as new technology takes over.
- Cheaper labour from countries such as India.
- Improved (and cheaper) transport and communication means that products do not need to be manufactured near to the market.
- Ending of government incentives to encourage new industries.
- Modernisation of plants in order to compete can result in redundancies.
- Rationalisation of company leading to overseas plants being closed.

(ii) Again, answers will depend on the area chosen but will include effects such as:
- Associated service and supply industries close.
- Cycle of economic decline sets in.
- Depopulation, particularly amongst young people and young families.
- Leading to further service closures.
- Areas in decline find it difficult to attract new investment as area becomes run down.
- Rise in cases of depression.
- Rise in crime rates in area.

GEOGRAPHY HIGHER ENVIRONMENTAL INTERACTIONS 2010

Question 1 – Rural Land Resources

Well annotated diagrams will be awarded full credit.

(a) Candidates should refer to the process of coastal erosion and deposition within their answer, ie hydraulic action, abrasion, solution, attrition and wave movement up/down beaches with longshore drift.

A typical answer for a cave/arch/stack may include:
Caves are most likely to occur where the coastline consists of hard rock and is attacked by prolonged wave attack along a line of weakness such as a joint or fault in the rock. The waves will attack the line of weakness by abrasion, hydraulic action or solution. Over time, horizontal erosion of the cave may cut through the headland to the other side, and form an arch. Very occasionally a blowhole will be created within the cave where compressed air is pushed upwards by the power of the waves and vertical erosion occurs. Continued erosion of the foot of the arch may eventually cause the roof to collapse leaving a stack, isolated from the cliff. This in turn will be eroded yet further to leave a stump.

(b) Candidates must discuss at least three land uses to achieve full marks.
Ideally candidates should identify the specific feature of the landscape and then go on to explain the opportunity it provides. Responses will vary according to the area chosen but opportunities might include:
Social – tourism, recreation, nature conservation.
Environmental – farming, forestry, energy generation (wind/waves/tidal), quarrying.
Industry – ports, oil industry.

(c) Candidates should be able to discuss both sides of the argument in this development, to achieve full marks.

Advantages/People arguing for the plan	Disadvantages/People arguing against the plan
Jobs will be created during and after the construction.	Labour force – probably not local people, strain on local services.
The golf course will boost visitor numbers and bring money into the economy of the local area.	This will damage the mobile sand dunes and associated wildlife. Loss of habitat/biodiversity.
The new houses will attract people to move here and boost local services like schools.	Locals/wildlife enthusiasts will no longer have access to the beach/sand dunes on the estate.
Local businesses will benefit by supplying the hotel ie taxi companies, food/farm contracts.	Car parks/roads and buildings associated with the development will cause visual pollution.

(d) To gain full marks candidates must comment on the effectiveness of their solutions/measures taken to resolve environmental conflicts.

Measures taken to resolve environmental conflicts might include:
- Traffic restrictions in more favoured areas.
- Reducing congestion on busy roads using a one-way system.
- Encouraging the use of minibuses.
- Seperating local and tourist traffic.
- Attempting to develop wider spread of 'honeypot' areas.
- Providing cheap local housing for inhabitants of area.
- Screening new buildings, car parks etc behind deciduous trees and using only local stone for buildings.
- Better visitor education.

Question 2 – Rural Land Degradation

(a)
- **Rainsplash** – usually the first stage in the erosion process, the impact of raindrops on the surface of a soil causing the soil particles to be moved. On steeper slopes they move further downhill. This means that resettled sediment blocks soil pores resulting in surface crusting and lower infiltration.
- **Sheet erosion** – the removal of a thin layer of surface soil which has already been disturbed by rainsplash, accounts for large volumes of soil loss, rarely flows for more than a few metres before concentrating into rills. Typically results in the loss of the finest soil particles which usually contain the nutrients and organic matter.
- **Rill erosion** – small eroded channels, only centimetres (up to about 30cm) deep and not permanent features, often obliterated by the next rainstorm, or develop into gullies.
- **Gully erosion** – steep sided water channels, several metres deep which can cut deeply into the soil after storms and are often permanent. Rain water running into the gully scours the sides or undercuts the head wall which results in the gully migrating. Widening of gully sides can occur by undercutting or slumping.

(b) Human causes of land degradation will vary according to the location chosen but may include reasons such as:

North America
For the **Dust Bowl**:
- Use of techniques better suited to the moister eastern states.
- Monoculture, especially of wheat or demanding crops (cotton), depleted the soil of moisture and nutrients.
- Deep ploughing of fragile soils (previously these had been held in place by natural grasslands).
- Marginal land ploughed – particularly in wet years – leaving them in a fragile condition in dry years.
- Ploughing downslope creating opportunities for rill erosion.
- Farm sizes being too small so forcing farmers to overcrop – particularly when prices were low and therefore income was low.

For the **Tennessee Valley**:
- Much of the area was cleared of its trees – this opened up the soil surface to erosion.
- Mining and farming also cleared the natural vegetation and led to soil erosion.
- The farmers cultivated steep slopes which were ploughed up and down the slope.
- Overcropping had already weakened the soil.
- The eroded soil was dumped in rivers and this caused them to flood.
- A lack of fertilizer caused the soil to lose its structure and become vulnerable to erosion.

For **Africa, north of the equator,** mention might be made of overgrazing, overcropping, deforestation, monoculture, farming cash crops.
- Allowing more grazing than the pasture can support (eg in West Africa) where herd size is a status symbol.
- Allowing the soil to be stripped bare leaving it vulnerable to erosion.
- Increased population density caused by falling death rates leading to overcultivation.

- Deforestation for firewood/building.
- Bush fires to clear land for farming.
- In some places peasant farmers have had to farm marginal land due to the best land being used for cash crops (eg in parts of Sudan).
- The drought may have caused nomads to move into villages where the land may now be over-cultivated (eg in Burkina Faso).

For the **Amazon Basin,** answers will be based on deforestation:
- Deforestation – for eg ranching/mineral extraction/logging/road building/poor peasant farmers.
- Loss of protective cover of trees due to deforestation.
- This allows heavy tropical rainfall to erode the soil.
- Exposure to increased sunlight due to deforestation leads to the soil baking and becoming useless.
- The loss of the root system which previously bound the soil together.
- Deforestation also leads to increased leaching of the soil rendering it useless in addition to erosion.
- The impact of ranching: forest cleared, used for a few years until grass fails – move and clear a new stretch of forest and continue the process.

(c) For **Africa, north of the equator** descriptions may include:
- Crop failures and the resulting malnutrition leading to famine eg Sudan, Ethiopia and much of the Sahel.
- Southward migration on a large scale – usually into shanties on the edge of the major cities.
- The collapse of the nomadic way of life due to the lack of grazing and water.
- Many nomads forced to settle in villages – with a consequent increase in pressure on the surrounding land.
- The breakdown of the settled farmer/nomad relationship in places like Yatenga province in Northern Burkina Faso.
- Disease and illness can become endemic.
- Conflict within countries as people move and re-settle.
- Countries increasingly rely on international aid.

For the **Amazon Basin** answers may include:
- Destruction of the way of life of the indigenous people eg clashes between the Yanomami and incomers.
- Destruction of the formerly sustainable development eg rubber tappers and Brazil Nut collectors.
- Clashes between various competing groups eg the violent death of Chico Mendez allegedly at the behest of ranchers.
- Reduction of fallow period leading to reduced yields with obvious consequences for the dependent population.
- Creation of reservations for indigenous people.
- Increase in 'western' diseases.
- Increase in alcoholism amongst indigenous population.
- People have been displaced and forced into crowded cities ending up living in favelas.

(d) Answers should be able to give reasonably detailed information about farming methods, and must include some explanation of these methods, for example:
Shelter belts – on low lying land affected by strong winds shelter belts are rows of trees grown across the direction of the prevailing wind. They act as a barrier to slow down winds and protect the soil. The taller and more complete the barrier of trees the more effective the shelter.

Other farming methods might include:
- Crop rotation.
- Diversification of farming types.
- Keeping land under grass or fallow.
- Trash farming/stubble mulching.
- Replanting shelter belts.
- Strip cultivation and intercropping.

- Increased irrigation.
- Soil banks by keeping soils under grass rather than ploughing.
- Diversification by farmers into recreation.
- Contour ploughing.
- Terracing.
- Use of natural fertilisers.

Question 3 – River Basin Management

(a) (i) Description and explanation of pattern of river flow might include:
- **Description** – very irregular flow from month to month from 2,000 cumecs in Jan/Feb to 13,000 cumecs in Aug/Sept.
- Similar pattern from year to year with peaks and troughs at the same time each year.
- **Explanation** should refer to the fact that troughs relate to dry months from Dec-Mar while peaks occur after heavy rainfall of Jun/Jul/Aug. Rainfall monthly figures indicate monsoon rains.
- Discharge increasing before heavy rainfall suggests discharge is fuelled by snow melt from surrounding mountains shown on reference map Q3A.

(ii) Description and explanation of need for water management might include:
- Reference map Q3A indicates that the Irrawaddy River has many tributaries and the river basin has a very high drainage density leading to unpredictability of river flow which is dependent on when and how quickly snow melts in surrounding mountain areas.
- Rapidly increasing population in Myanmar gives increasing demand for water for domestic, power, industrial needs.
- Increasing demands from farmers for irrigation water to try and feed increasing population.
- Rainfall graph for Myitsore indicates seasonal nature of rainfall – extremely dry from November to April but huge monthly figures for June/July/August – leading to flooding and also run-off of water that could be stored and used in dry months.
- Temperature graph for Myitsore indicates hot temperatures throughout the year leading to very high evaporation rates. Monthly temperatures peak at 35°C.
- Reference diagram Q3A indicates that there is a need to regulate flow of river to prevent flooding during peak discharge and to keep water level high enough for navigation in dry months.

(b) Physical factors might include:
- Geologically stable area away from earthquake zones/fault lines.
- Solid rock foundations for weight of dam.
- Narrow valley cross-section to reduce dam length.
- Large, deep valley to flood behind dam to maximise amount of water storage.
- Lack of permeability in rock below and around reservoir to prevent seepage.
- Low evaporation rates.
- Large catchment area above dam to provide reliable water supply.

(c) Answers should be authentic for the chosen river basin. Candidates must refer to all 6 sections for full marks.

Answers will depend on the river basin chosen. However, for the Colorado River they might include:

Social benefits:
- Fresh water supply for growing desert cities eg Phoenix.

- Better standard of living in hot, dry climate with air conditioning, swimming pools, landscaping etc.
- Areas at reservoirs, eg Lake Mead, give opportunities for tourism, water sports, fishing etc.
- Regulation of river greatly improves flood control on river.

Social adverse consequences:
- People had to be moved off their land as valley areas were flooded.
- Loss of burial sites and other Native American sacred areas.
- Disagreements between states and countries with regard to allocation of water from river.

Economic benefits:
- Cheap HEP attracted industries eg electronics to take advantage of the area's cheap land and low taxes.
- Benefited tourist industry with reliable water supply – attractions like the Grand Canyon, gambling in Las Vegas, Hoover Dam etc.
- Expansion of irrigated land led to agribusiness-style farming.

Economic adverse consequences:
- Huge cost of building the dams eg Central Arizona Project cost $6 billion.
- High cost of maintaining dams, power plants and irrigation channels.
- Subsidised water for farmers has led to water wastage and the growing of crops that could be produced cheaper elsewhere.

Environmental benefits:
- Reservoirs provide sanctuaries for waterfowl and wading birds like the blue heron.
- The National Recreation Area around Lake Mead has more than 250 species of birds.
- Reliable seasonal water flow for plant and animal life.

Environmental adverse consequences:
- Water in river and on farmland becomes saline with high evaporation rates – farmers downstream have to switch to more salt-tolerant crops.
- Change in river regime has caused the loss of many animal habitats eg the drying up of Colorado delta area where there used to be a great variety of birdlife.
- Huge amounts of water loss by seepage through the sandstone rocks around Lake Powell. Scenic attractions like the Rainbow Bridge are being affected by the high water levels in Lake Powell.

Question 4 – Urban Change and its Management

(a) Candidates should be able to identify the overall increase in number of megacities from 1975 to 2015.
They should recognise that the greatest area of growth is in developing countries rather than developed countries.
Within this they should identify patterns of increased growth in particular areas of the world eg India and China in Asia and USA and Brazil in the Americas.

(b) (i) Candidates should be able to demonstrate authentic knowledge of a city they have studied. Answers could include the following points:

Rural push:
- Low income from farming and related work.
- Lack of employment in manufacturing and service industry.
- Lack of education and low literacy levels.
- Poor health facilities and higher levels of disease, malnutrition etc.
- Low quality of life, poor sanitation, lack of electricity.
- Poor quality of infrastructure.
- Resettlement, civil unrest, environmental degradation.

Urban pull:
- Industrial employment, both manufacturing and service.
- Informal opportunities for employment.
- Increased income.
- Better housing, education, health facilities.
- Improved infrastructure.
- 'Bright lights' ambitions.

(ii) Candidates should relate socio-economic and environmental problems to their chosen city.
- Impoverished and overcrowded areas of the city which lack many public utilities and amenities of water supply, electricity and sewerage.
- Semi-urban peripheral districts with poor housing quality and poor economic opportunities. Squatter settlements are located on steep upland areas. Areas lack basic services eg schools, piped water and hospitals.
- High incidence of disease.
- High rates of unemployment and growth of 'grey' economy and black market.
- Crime, drugs and prostitution.
- Problems of waste disposal include open sewers, toxic industrial waste contaminating water supply, lack of refuse collection and landfill sites for solid waste.
- Air pollution caused by chronic traffic congestion and industrial emissions.

(c) (i) Reasons for urban sprawl might include:
- Growth of population.
- Growth of suburban housing both high quality private and low cost, council estates.
- Cheaper land prices on outskirts.
- Development of shopping malls, industrial estates and retail parks.
- Growth of leisure facilities eg golf courses, new football stadia.
- Need for motorway and by-pass developments.
- Increased level of commuting to suburban areas and villages with more attractive environments.
- Negative aspects of the city eg pollution, congestion, land prices, house prices, levels of social problems like crime.
- Increased number of single person households.

(ii) Candidates could look at problems at the edge of the city as well as within the inner urban areas. Problems might include:
- Urban sprawl using up recreational and farm land.
- Urban sprawl threatening wildlife habitats and removing clean air lungs and open land.
- Increased commuting leading to traffic congestion and increasing levels of air pollution.
- Buildings and services in inner urban areas not being used or becoming run down or derelict eg housing, schools, factories and shopping areas.

(iii) Candidates require to select one problem and identify how their chosen city has dealt with the problem. For example if the candidate had chosen traffic congestion the following solutions could be developed.
- Policies to reduce cars eg car sharing, high occupancy vehicle lanes, new car tax charges, congestion charges, cycle routes.
- Promotion of improved public transport, including lower pricing, integrated transit systems.
- Park and Ride schemes.

- Mass transit systems using fixed routes eg metro lines and tramways.
- Changing road systems eg flexi-time travel, tidal flows, coordinating traffic lights, bus lanes.

Question 5 – European Regional Inequalities

(a) (i)
- Credit should be awarded for candidates noting that Convergence Regions are found in Europe's peripheral areas, notably in eastern periphery countries with former centrally planned economies eg Bulgaria, Hungary, Baltic states etc.
- Also southern European peripheral areas such as southern Italy and much of Greece, Spain and Portugal.
- In the UK, Cornwall and western Wales have this status.
- No Convergence Regions found in most of northern, central and western areas of the EU.

(ii) EU measures could include:
- Cohesion Fund – aimed at member states whose Gross National Product (GNP) per inhabitant is less than 90% of the EU average. It serves to reduce their economic and social shortfall, as well as to stabilise their economy. The Cohesion Fund finances activities such as trans-European transport networks and also projects related to energy or transport as long as they clearly present a benefit to the environment.
- The European Regional Development Fund (ERDF) aims to strengthen economic and social cohesion in the EU by correcting imbalances between its regions by financing technical assistance measures, improvements to local infrastructure etc.
- The task of the European Investment Bank (EIB), the EU's financing institution, is to contribute towards the integration, balance, development and economic and social cohesion of all the member states.
- The European Social Fund (ESF) set out to improve employment and job opportunities in the EU through lifelong learning schemes and providing access and employment for job seekers, the unemployed, women and migrants. It supports actions to socially integrate disadvantaged people, combating discrimination in the job market.

For full marks, candidates will require to illustrate points made with some well-chosen examples and statistics.

(b) (i) Candidates should use some form of comparative statements covering all four indicators to get full marks.

The four indicators given all identify a similar pattern identifying regional inequalities within the UK –East England, SE and SW England and London generally fare better than Wales, the West Midlands and areas further north.
- Population change – apart from Northern Ireland, regions with the highest increase in population are in the south of England. No growth or decrease in Scotland, NE and NW England.
- Average house prices – highest in London and SE England, while Scotland, northern regions of England and Wales have figures well below the UK average.
- Gross Disposable Household Income – very similar to house prices although Yorks and Humberside, East Midlands and Scotland fare slightly better.
- Working Age Population with no qualifications (%) – very low in southern England, again high in Wales, northern England and especially Northern Ireland. Scotland same as UK average.

(ii) The UK's regional inequalities stem from a combination of the physical differences between the higher and steeper land to the north and west of the UK compared with the lower and more gently sloping land to the south and east coupled with the remoteness of the north-west compared to the proximity of the south-east to the 'core' of the EU. Candidates may justifiably stress the positive and negative aspects of different regions.
- Physical factors might mention advantages/ problems such as relief, rock types, climate and water supply, soil fertility and erosion.
- Human factors might mention decline of traditional heavy industries, growth areas of new lighter industries and hi-tech industries, out-migration from north and differences in accessibility related to communications and remoteness.

(iii) UK national government help could include:
- UK government identifies 'Assisted Areas' eligible for regional selective assistance in line with EU moves to redistribute most of aid budget to poorer areas. Assisted areas include the whole of Northern Ireland, Cornwall and the Scilly Isles, West Wales and the Valleys and the Scottish Highlands and Islands.
- There are also economic development agencies in the four countries of the UK which aim to attract investment and help new and existing businesses compete nationally and internationally.
- The Welsh Assembly Government's department of Economy and Transport in Wales.
- Scottish Enterprise and Highlands and Islands Enterprise in Scotland.
- Invest Northern Ireland in NI and The Regional Development Agencies (RDAs) in England.

Question 6 – Development and Health

(a) Candidates should be able to refer to:
- Oil rich countries such as Saudi Arabia, Brunei; well-off countries like Malaysia which can export primary products such as hardwoods, rubber, palm oil and tin.
- Poor Sahelian countries like Mali, Chad and Burkina Faso which are landlocked, lack resources, have poor quality farmland, high levels of disease.
- Newly Industrialised Countries eg South Korea, Taiwan have high GNPs due to steel-making, shipbuilding, car manufacturing, clothing etc. Countries with entrepreneurial skills and low labour costs.
- Large countries eg Brazil with a variety of opportunities ranging from resources in Amazonia to tourism in the South East around Rio.
- Tourist destinations eg Sri Lanka, Thailand, Caribbean islands like Barbados, earn foreign currency and improve living standards and create new job opportunities.
- Countries which suffer natural disasters which restrict development and cause massive damage to infrastructure. Examples include drought in Ethiopia, floods/cyclones in Bangladesh, hurricanes in Caribbean and tsunamis in Indonesia, earthquakes in China.
- Mountainous countries eg Tibet, Afghanistan which restrict communications and farming.
- Areas of political instability which divert aid and resources away from areas of need. Examples include civil war in Sudan, large scale conflicts in Afghanistan and Iraq, corruption, mismanagement and need for regime change in Zimbabwe.

(b) Candidates can make use of resources to suggest the following factors which may lead to low life expectancy.
- Chad is landlocked restricting trade, income from imports/exports and reducing money available to improve quality of life.
- Low GDP means Chad will struggle to provide services such as hospitals/clean water/sanitation which will increase ill health.
- High infant mortality will reduce average life expectancy.
- Low quantities of farmland and irrigated land mean crop production will be less than required to feed population leading to malnutrition and ill health.
- Low literacy levels imply poor education in areas of hygiene/birth control/disease control.
- Inhospitable areas eg desert, uplands means living conditions are very harsh
- High fertility rates mean large families with not enough food and resources to go round.
- High levels of water borne diseases reduce life expectancy.

(c) (i) Answers will depend on the disease chosen but for Malaria might include:

Physical factors:
- Hot wet climates such as those experienced in the tropical rainforests or monsoon areas of the world.
- Temperatures of between 15°C and 40°C.
- Areas of shade in which the mosquito can digest human blood.
- Migration.
- Not completing course of drugs.

Human factors:
- Suitable breeding habitat for the female anopheles mosquito – areas of stagnant water such as reservoirs, ponds, irrigation channels.
- Nearby settlements to provide a 'blood reservoir'.
- Areas of bad sanitation, poor irrigation or drainage.
- Exposure of bare skin.
- Migration.
- Not completing course of drugs.

(ii) Measures taken to combat malaria may include:
Trying to eradicate the mosquitoes:
- Insecticides eg DDT.
- Newer insecticides such as Malathion
- Mustard seed 'bombing' – become wet and sticky and drag mosquito larvae under the water drowning them.
- Egg-white sprayed on water – suffocates larvae by clogging up their breathing tubes.
- BTI bacteria grown in coconuts. Fermented coconuts are, after a few days, broken open and thrown into mosquito-infested ponds. The larvae eat the bacteria and have their stomach linings destroyed!
- Larvae eating fish.
- Drainage of swamps.

Treating those suffering from malaria:
- Drugs like chloroquin, larium and malarone.
- Quinghaosu extract from the artemesian plant – a traditional Chinese cure.
- Continued search for a vaccine – not available as yet.
- Education programmes in –
  - the use of insect repellents eg Autan
  - covering the skin at dusk when the mosquitoes are most active
  - sleeping under an insecticide treated mosquito net
  - mesh coverings over windows/door openings.
- WHO 'Roll back malaria' campaign.
- The Bill and Melinda Gates Foundation.

(iii) The benefits of controlling the disease on a developing country might include:
- Saving money on health, medicines, doctors, drugs etc.
- Reduction in the national debt.
- Healthier workforce and increased productivity.
- Longer life expectancy and decreased infant mortality rates.
- Scarce financial resources could be spent on other areas such as education or housing.
- More tourists/foreign investment may be attracted if there was less risk of disease – leading to more job opportunities, foreign currency earnings, increased prosperity.

GEOGRAPHY HIGHER
PHYSICAL AND HUMAN ENVIRONMENTS
2011

SECTION A

Question 1 – Atmosphere

(a) Maritime Tropical (mT)
- Origin – Atlantic ocean/Gulf of Guinea, in tropical latitudes
- Weather characteristics – hot, high humidity, warm
  Nature – unstable

Continental Tropical (cT)
- Origin – Sahara Desert, in tropical latitudes
- Weather characteristics – hot/very hot, dry, low humidity, warm
  Nature – stable, poor visibility

(b) Description should highlight the marked contrast in precipitation totals, seasonal distribution and number of days between a very dry north (Gao with only 200 mm in a hot desert climate in Mali) and a much wetter south (Abidjan with 1700 mm in a tropical rainforest climate in the Ivory Coast).

Bobo-Dioulasso in Burkina Faso in central West Africa has an 'in-between' amount of both rain days and total annual precipitation (1000 mm in a Savannah climate).

Candidates should also refer to the variation in rain days and seasonal distribution for each station. Gao with a limited amount of precipitation in summer, Bobo-Doiulasso with a clear wet season/dry season regime and Abidjan with a 'twin-peak' regime with a major peak in June and a smaller peak in October/November.

Explanation should focus on the role of the ITCZ and the movement of the Maritime Tropical and Continental Tropical air masses over the course of the year. For example, Abidjan, on the Gulf of Guinea coast, is influenced by hot, humid mT air for most of the year, accounting for its higher total annual precipitation and greater number of rain days. The twin precipitation peaks can be attributed to the ITCZ moving northwards in the early part of the year and then southwards later in the year in line with the thermal equator/overhead sun.

Gao, on the other hand, is under the influence of hot, dry cT air for most of the year and therefore has far fewer rain days and a very low total annual precipitation figure as it lies well to the north of the ITCZ for most of the year. Bobo-Dioulasso again is in an 'in-between' position, getting more rain days and heavy summer precipitation from June-August when the ITCZ is furthest north.

Question 2 – Biosphere

(a) Climax vegetation is the final stage in the development of the natural vegetation of a locality or region when the composition of the plant community is relatively stable and in equilibrium with the existing environmental conditions. This is normally determined by climate or soil. These are self-sustaining ecosystems.

Candidates should be credited for being able to demonstrate knowledge of the evolution of plant life from early colonisation by pioneer species then, by succession, to the ultimate vegetation climax. Appropriate examples could also be given credit eg oak-ash forest in a cool temperature climate such as exists over much of Britain or Scots pine-birch forest in colder, wetter and less fertile Highland environments.

(b) **Strandline (Sea Sandwort, Sea Rocket, Saltwort, Sea Twitch)**
These are all salt tolerant (halophytic) species and can withstand the desiccating effects of the sand and the wind. Some can even withstand periodic immersion in sea water. There is a high pH here (alkaline conditions) due to the presence of sea shells. The presence of these plants leads to further deposition of sand and the establishment of less hardy species.

Embryo Dune (Sea/Sand Couch, Lyme Grass, Frosted Orache, Sea Rocket)
These dune pioneer species grow side (lateral) roots and underground stems (rhizomes) which bind the sand together. These grassy plants can also tolerate occasional immersion in sea water. Some species on the strandline are also found in the embryo dunes.

Fore Dune (Sea Bindweed, Sea Holly, Sand Sedge, Marram Grass)
A slightly higher humus content (from decayed plants), and lower salt content (further from the sea) allows these species to further stabilise the dune and allow the establishment of Marram Grass which becomes a key plant in the build up of the dune.

Yellow Dune (Marram Grass, Sand Fescue, Sand Sedge, Sea Bindweed, Ragwort)
Both the humus content and the acidity of the soil have increased at this location. Marram can align itself with the prevailing wind and curl its leaves to reduce moisture loss; it can also survive being buried by the shifting sand of the dune. As sand deposition increases the Marram responds by more rapid rhizome growth (up to 1 metre a year). It is xerophytic, and so is better able to survive the dry conditions of the dune. It also has long roots which help to bind deposited sand and anchor it into the dunes as well as access water supplies some distance below. All these factors allow it to become the dominant species on the Yellow Dune.

Grey Dunes (Sand Sedge, Sand Fescue, Bird's Foot Trefoil, Heather, Sea Buckhorn, Grey Lichens eg Cladonia species)
As a result of an increase in organic content (humus), greater shelter and a damper soil, a wider range of plants can thrive here. Marram dies back (contributing humus) to be replaced by other grasses. As a result of leaching and the build up of humus the soil is considerably more acidic allowing more plant species to flourish.

Slacks (reeds, rushes, cotton grass, flag iris, alders and small willow trees)
In the wetter slacks, close to the water table, several water loving (hydrophytic) species may survive.

Climax (Heather, trees such as Birch, Pine or Spruce)
In some areas heathland may dominate with a range of heathers being prominent. Eventually trees such as Birch, Pine or Spruce could establish a hold. In the shell rich areas of the Western Isles, Machair may develop.

Question 3 – Rural Geography

(i) For **shifting cultivation** the main characteristics of the landscape might include:
- clearings are made in the rainforest by cutting down and burning trees
- largest trees and some fruit-bearing trees may be left for protection/food (some are too difficult to remove)
- the 'cultivation' part refers to the practice of growing crops (manioc/cassava, yams ...) in the clearing, using ash from the tree burning as fertiliser

- the 'shifting' part refers to the practice of moving to another clearing as the soil becomes exhausted and crop yields fall
- low population density due to large area of land needed
- settlements could be fixed (and rotational clearings made around them) or the housing may also be abandoned and left to biodegrade before the tribe returns to the area.

(ii) For **shifting cultivation** the main changes might include:
- loss of traditional tribal land due to cattle ranching, mineral extraction, logging, HEP development
- change in land use with set reservations/settlements, National Parks and conservation areas
- climate change with increasing unpredictability of drought/flood cycles.

For **shifting cultivation** the main impacts might include:
- population movement into inaccessible areas which are often less fertile
- rural depopulation with shanty town growth in larger urban areas
- contact with Western culture can bring diseases, alcohol/drug misuse
- population densities increase in remaining areas, putting more strain on limited land and a shorter fallow period.
- decreasing soil fertility and output per hectare
- soil erosion can take place with the soil choking the rivers reducing fish/wildlife in the area/impact on diet.
- pollution from other land users ie mercury used in gold extraction can impact on the health of the locals
- the impact of global warming on biodiversity and medicinal cures.

Question 4 – Industry

(a) Reasons for location of industry in Swansea may include:

Physical factors:
- Area A on the outskirts of town with flat land, and room for expansion, Area B is on flat floodplain of Afon Tawe.
- Both have flat land for easy construction of industrial buildings.

Human factors:
- Proximity to local market in South Wales.
- Access to docklands for import and export and via Afon Tawe.
- Close to motorways for easy access of materials and finished goods,a and branch line 683970
- Proximity to local labour force.
- Edge of town – cheaper land.
- Close to universities for skilled graduates and research facilities.
- Close to other modern industries that may supply components or share resources.
- Swansea airport for visiting executives, or transporting light products (5691).
- Pleasant working environment.

(b) Answers will vary, depending on industrial concentration chosen.

Explanations may include:
- Creation of Enterprise Zones.
- Welsh Development Agency.
- Creation of new town (Cwmbran).
- Rent free accommodation.
- Grants.
- Retraining schemes.
- Relocation of industry (government offices eg DVLA, and foreign businesses eg Sony).
- Tax incentives.
- Road and infrastructure projects (M4).
- Environmental improvement schemes.
- Objective 1 funding.

SECTION B
Question 5 – Lithosphere

(a) Descriptions could include:
- Cliffs eg Newton Cliff (GR 600870).
- Headland eg Pwlldu Head (GR 570863).
- Caves eg Mitchin Hole Cave (GR 555869).
- Blow Holes eg Bacon Hole (GR 561868).
- Shore (wave-cut) platform (GR 615869).
- Stack eg Mumbles Head (GR 636871).
- Skerries (stack or stump) eg Rothers Sker (GR 612869).
- Bay (GR592874) – Caswell Bay

(b) **A quality diagram could achieve full marks.**
Candidates should refer to the processes of coastal erosion ie hydraulic action, abrasion, solution and attrition. A typical answer may include:
Caves are most likely to occur where the coastline consists of hard rock and is attacked by prolonged wave attack along a line of weakness such as a joint or fault. The waves attack the weakness by abrasion, hydraulic action or solution. Over time, horizontal erosion of the cave may cut through the headland to form an arch. Continued erosion of the foot of the area may eventually cause the roof to collapse leaving a stack, isolated from the cliff.

Question 6 – Hydrosphere

(a) Answers should refer to the four elements in a drainage basin:
- input: precipitation
- storage: surface storage eg lakes, soil moisture, ground water, interception
- transfers: surface run off eg tributaries, throughflow, groundwater flow, infiltration, throughfall, percolation, stem flow
- outputs: transpiration, evaporation, surface run-off (rivers)

(b) Answers should identify various parts of the river level graph.
- Steady river level (under 0.4m) until 03:00 hours due to an initial lack of rain and then small amounts of rain at 05:00 hours (0.5mm) and 06:00 hours (0.8mm) infiltrate the soil (after interception by vegetation) and the river level starts to increase slowly.
- The river level continues to rise at a steady rate from 07:00 hours to 10:00 hours, due to the increase in rainfall totals and duration. The heavier rain is filling up storages in the soil because of throughflow and groundwater. The soil is now saturated, so water runs off the land and enters the river quickly leading to a potential flood situation.
- The peak rainfall occurs at 08:00 hours (6.2mm) and the peak river level occurs at 18:00 hours (0.7m). This is a basin lag time of approximately 10 hours. This could be accounted for by vegetation cover, or by reference to geology or soil infiltration rates.
- From 14:00 hours to the end of the graph the rainfall declines and stops at 18:00 hours. The recession limb falls back towards base level as the supply of water is reduced.

SECTION C

Question 7 – Urban Geography

(a) Answers will vary according to the city studied but may include reference to:

Site
- Flat land.
- Inside a large river meander.
- Early functions eg religious, defensive, trading site.
- Raw materials.
- Lowest bridging point.

Situation
- Easily accessible to major settlements.
- Accessible to ports.
- Major route focus.
- Accessible to airports.

(b) Answers will vary according to the city studied but for the inner city changes might include:

Description of the changes:
- Population reduced as people moved out of the area.
- Redevelopment of housing and the area.
- Construction of council houses and flats on cleared areas of the old inner city.
- Demolition of some terraces, improvements to others (indoor toilets etc).
- Environmental improvements eg parks, community centres, leisure centres.
- Relocation of industry and the closure of industry.
- Changes in the ethnicity of residents.

Explanation of the changes:
- Area was overcrowded.
- Housing was in a very poor condition and amenities were poor.
- High levels of unemployment, poverty and social problems such as crime.
- The environment was very poor and inward investment was difficult to attract.
- Industry declined as manufacturing moved to countries with lower labour costs.
- Industry has moved to modernised areas on the edge of the city in new custom-built units.
- Congested traffic and air pollution put off potential investors.
- Movement out to the suburbs due to the desire for a better quality environment.

Question 8 – Population Geography

(a) Between censuses there is compulsory registration of births, deaths and marriages by the General Register Office for each part of the UK. The other main changes are brought about by emigration and immigration and the Home Office's UK Border Agency records migration into the UK.

Also mini or sample censuses are carried out such as the 2009 Census Rehearsal for England and Wales as well as government sponsored sample surveys of population and social trends.

The UK official 2011 Census website identifies 6 areas of accurate population data for the targeting of taxpayers' money.
- Population numbers – to calculate grants for local authorities to plan eg schools and teacher numbers.
- Health – to know the age and socio-economic make-up of the population to allocate health and social services resources.
- Housing – to ascertain the need for new housing.
- Employment – to help government and businesses plan jobs and training policies.
- Transport – to identify where there is pressure on transport systems and for planning of roads and public transport.
- Ethnic Group – to identify the extent and nature of disadvantage in Britain.

(b) Difficulties affecting accurate population data collection in Developing Countries might include:
- Countries suffering from a continuing war situation such as Afghanistan.
- Illegal immigrants wishing to avoid detection eg Mexicans in southern California.
- The cost involved in carrying out a census is prohibitive to many Developing Countries – training enumerators, printing and distributing forms etc.
- The sheer size of some developing countries eg Indonesia with many islands spread over a large area.
- Suspicion of the use of data collected, eg China's one-child policy with many female births unrecorded.
- Countries with large numbers of migrants eg rural-urban migration into massive shanty towns eg Kibera in Nairobi, refugees from Rwanda in Burundi etc.
- Nomadic people such as the Tuareg in West Africa, shifting cultivators in Amazonia.
- Poor communication links eg mountain regions of Bolivia.
- Low levels of literacy and variety of languages spoken within a country eg India has 15 official languages.

GEOGRAPHY HIGHER ENVIRONMENTAL INTERACTIONS 2011

Question 1 – Rural Land Resources

(a) For an answer to achieve full marks, well annotated diagrams must be used. For full marks a minimum of two features must be described and explained, eg for a corrie points could include:
- Snow accumulates in north/east-facing hollow due to lack of melting.
- Successive layers of snow compress into ice/neve.
- Ice moves downhill under gravity.
- Freeze-thaw weathering occurs on backwall.
- Plucking steepens the backwall.
- Boulders embedded in ice grind away at bottom of the corrie.
- Abrasion carves out armchair-shaped depression due to rotational movement.
- Rate of erosion decreases at edge of corrie leaving a rock lip.

(b) Answers are expected to link these opportunities to the physical landscape, and answers must mention both social and economic opportunities for full marks.

Explanations can be developed from:

Social opportunities
- Mountaineering and hillwalking.
- Forest walks, picnic sites and orienteering courses.
- Sailing, fishing and other water sports.
- Nature conservation.

Economic opportunities
- Tourism and associated employment and profits.
- Development of hotels, bunkhouses and campsites.
- Hill sheep farming.
- Forestry plantations.
- HEP and water supply.
- Quarrying.

(c) Answers should be able to compare the popularity of parks based on analysis of the resource:
- Lake District is close to heavily populated areas, Merseyside, Yorkshire, Manchester.
- Snowdonia more remote in North-West Wales.
- Lake District is more accessible by motorway especially for short visits, M6, M74.
- Snowdonia is less accessible by motorway.
- Credit can also be given for candidates' knowledge of both parks, in terms of attractions.

(d) (i) Answers should be able to explain the environmental conflicts including:
- Traffic congestion especially on narrow rural roads and in car parks especially at peak holiday periods.
- Increased air and noise pollution.
- Increased holiday homes which leave rural areas empty during the week or off peak.
- Footpath erosion.
- Disruption to farms, damage to walls, disturbance to animals.
- Litter.
- Unsightly buildings including hotels, leisure complexes, caravan sites.
- Impact on lakes, bank erosion and diesel pollution due to water sports.

(ii)
- One way streets, bypasses, wardens, parking restrictions.
- Encourage use of public transport eg park and ride, minibus.
- Use of cycle paths, bridle ways, long distance paths.
- Use of permits to separate locals and visitors.

Question 2 – Rural Land Degradation

(a) The four main processes of erosion by water can be described as:
- Rainsplash – the impact of raindrops on the surface of a soil.
- Sheet wash – the removal of a thin layer of surface soil which has already been disturbed by rainsplash.
- Rill erosion – small eroded channels, only a few centimetres deep and not permanent features, often obliterated by the next rainstorm.
- Gully erosion – steep sided water channels, several metres deep which can cut deeply into the soil after storms and are often permanent.

The three main processes of wind erosion can be described as:
- Surface creep – the slow movement of larger (and heavier) particles across the land surface.
- Saltation – the bouncing along of lighter particles.
- Suspension – the lightest particles (dust) blown off ground for up to several hundred kilometres, dust storms.

(b) Answers should include the following descriptions and explanations from the resources:
- Niger is a landlocked, dry country – part of the Sahel Zone.
- Extreme range of temperatures on a daily basis.
- Seasonal rainfall concentrated from April to September.
- High temperatures coinciding with highest precipitation leading to high evaporation.
- Annual rainfall creates desert conditions.
- Variable annual rainfall from 1950 to 2010 with periods above average encouraging farming even in marginal areas, and periods below average leading to drought and degradation.
- Clear skies and strong direct sun can bake ground.

Candidate must relate aspects of climate to degradation caused by wind and water erosion.

(c) Answers should include explanations in the four areas of human activity outlined (Africa north of the Equator):
- Deforestation for firewood and farmland left soil exposed to erosion, removed root systems which would hold soil together, and removed shelter belts and wind breaks.
- Overgrazing exposed soil to winds by loss of vegetation cover, hooves break up soil making it susceptible to wind and water erosion, and in some cases compact the soil, especially near water holes.
- Overcropping means soil structure breaks up, with monoculture depleting nutrients, reduced fallow times meaning soil cannot rest or recover, marginal land eg slopes being used and becoming susceptible to wind and water erosion.
- Inappropriate farming techniques including monoculture, inappropriate ploughing eg deep ploughing of fragile soils, irrigation leading to salinisation, lack of organic fertilisers used.

(d) Soil conservation strategies might include:
- Crop rotation.
- Diversification of farming types.
- Keeping land under grass or fallow.
- Trash farming/stubble mulching.
- Replanting shelter belts.
- Strip cultivation and intercropping.
- Improved irrigation.

- Soil banks.
- Contour ploughing.
- Terracing.
- Use of natural fertilisers.
- Gully repair.
- Re-afforestation of slopes and marginal land.

Question 3 – River Basin Management

(a) Candidates may mention a range of reasons to explain the need for water management including:
- Very high rainfall in Borneo (tropical rainforest conditions).
- Flood control.
- Regulating flow and storage of water.
- Power supply for expanding cities and industry.
- Export of surplus electricity.
- Water for industrial purposes.
- Drinking water for increasing population.

(b) Physical factors might include:
- Solid foundations for a dam.
- Consideration of earthquake zones/fault lines.
- Narrow cross-section to reduce dam length.
- Large, deep valley to flood behind the dam.
- Lack of permeability in rock below reservoir.
- Sufficient water supply from catchment area.
- Low evaporation rates.
- Impact on the hydrological cycle.

(c) (i) Answers should be authentic for the chosen river basin. Answers will depend upon the basin chosen. However, some suggestions are outlined below:

Social:
- Greater population can be sustained with increased fresh water supply.
- Less disease and poor health due to better water supply/sanitation.
- Recreational opportunities/tourism on rivers and reservoirs.
- More widespread availability of electricity (and therefore modern technology/development).
- Floods could be avoided.

Economic:
- Improved farming outputs with possible surplus for sale.
- HEP – industrial development creating job opportunities eg Borneo could export surplus electricity to Indonesia and transfer to mainland Malaysia.
- Water (and power) for industry eg a proposed aluminium smelter in Borneo linked to the Bakun Dam.
- Navigation opportunities.

Environmental:
- Increased fresh water supply improves sanitation and health.
- Scenic improvement.

(ii) Answers will depend on the dam studied but for the Bakun Dam, answers may include:
- Threatened wildlife – endangered species may vanish, loss of tourism revenue, loss of medicinal plants/drugs still to be discovered.
- Forest cover – deforestation from reservoir/dam construction and increased access to remote areas with the resulting impact on local climate and global warming, silting up of rivers/soil erosion.
- Changing landscapes – reduced supply of timber for world trade and the increased planting of palm oil for biofuels/oil exports replacing the rainforest and resultant loss of biodiversity as access improves.
- People – relocation of indigenous tribes, loss of land/traditional way of life, spread of water-borne diseases from reservoirs.

The candidate may also include facts from the reference diagrams:
- There is already a surplus of electricity.
- The transmission line to transfer more electricity to the Malaysian mainland is only 'proposed'.

Other factors such as cost or political factors could be made relevant by the candidate.

Question 4 – Urban Change and its Management

(a) Answers will depend on the Developed World city chosen, but for the UK answers might suggest favourable locations for cities eg:
- Coastal locations – for trade with Europe/America (London/Glasgow), fishing industry (Aberdeen) and ship-building industry (Glasgow/Belfast).
- Natural routeways/rivers/canals – for communication, trade in raw materials (Leeds, Manchester, Sheffield, Birmingham).
- Access to raw materials – for coal, iron ore and limestone for the iron and steel industry (Glasgow/Sheffield).
- Historical/political factors in location of capital/primate cities – royal residences, parliaments etc (London, Edinburgh, Cardiff).

Whereas negative locations may also be included eg:
- Mountainous/upland areas – Highlands, Pennines.
- Inaccessible/marshy areas – Islands, Fens

(b) Candidates should note the advantages for the residents of the East End of Glasgow using the statistics from the diagrams/table.

Advantages could include:
- Improved communications – motorway/road extensions and improvements.
- Jobs – in the various venues before, during and after the event (construction, catering, transport etc).
- Training – the promise of skills training in various volunteer roles during the competitions that could then lead to permanent jobs.
- Social – attending/participating in the competitions.

Disadvantages could include:
- Disruption and pollution from the construction process in producing the 30% of the venues/facilities still to be built.
- Loss of land, access, community spirit/cohesion that new roads/motorways can lead to, particularly with the M74 extension cutting through a densely populated area.
- Some landowners/developers may have had to sell their property under compulsory purchase orders that may have led them to lose potential profits.

(c) Traffic congestion in a Developed World city. For Glasgow, candidates might suggest:
- An urban core developed in the pre-car era with medieval/Victorian sections unsuited for modern traffic (narrow, cobbled, grid-iron pattern with many junctions).
- Increased commuting from dormitory towns and villages converging on a few main arteries (Paisley Road West, Great Western Road, M8, Kilmarnock/Ayr Road).
- Major roads converging to cross the River Clyde (Clyde Tunnel, Kingston Bridge).
- Glasgow is a growing tourist/shopping centre attracting coach tours and shoppers from a larger hinterland.
- More stringent traffic regulations in and around the CBD with a shortage of cheaper gap site car parking facilities.

- Growing car ownership and school run traffic extending and expanding the 'rush hours'.
- Increased road haulage/deliveries in larger vehicles with a resulting need for more road maintenance.

(d) (i) The problems should be relevant to the candidate's chosen city and might include:
- Chaotic urban infrastructure eg incomplete water and sewerage supplies and connections leading to the spread of disease.
- Unemployment/underemployment
  – Growth of the 'grey' economy and black market
  – Drugs, crime, racketeering and prostitution are common and often involve a greater % of the population than a city in developed country
  – Poor wages for unskilled jobs partly due to the huge supply of labour available.
- Lack of services, schools and hospitals.
- Difficulties in encouraging city/public employees to work in the 'shanty' areas.
- Chronic traffic congestion and associated high levels of atmospheric pollution
  – Proliferation of 'informal' city transport (having both advantages and disadvantages.)
- Continued growth of 'shanty towns' in a range of locations in and around the city
  – 'Natural' disasters such as landslides resulting from the inappropriate building techniques and methods on fragile or unsafe land.

(ii) Again the methods used to tackle the problems should be related to the candidate's chosen city.
Candidates could offer a number of 'generic' solutions:
- An increase in the empowerment of the local people often with the aid of charity/church groups which provide advice/counsel/ lobbying facilities for the poorest elements of the population.
- Local council plans to improve basic infrastructure, including provision of water/sewerage to established 'shanties'.
- Improvements in the standard of basic education.
- The provision of hardware/utilities with the local populace providing the skill/effort to install these ie the 'basic shell' of housing being provided.

Specific solutions related to the candidate's chosen city are also wanted, these could include government drives to demolish squatter settlements and re-house the residents in new housing schemes.

Question 5 – European Regional Inequalities

(a) Countries may wish to become members of the European Union for the following reasons:
- Removing trade barriers to boost growth and create jobs.
- Tackling climate change and promoting energy security.
- Improving standards and rights for consumers.
- Fighting international crime and illegal immigration.
- Bringing peace and stability to Europe by working with its neighbours.
- Giving Europe a more powerful voice in the world.
- Securing food supplies and essential raw materials.
- Improving standards of living in the member states.

Specific EU measures to aid development include:
- European Regional Development Fund (ERDF) which provides a wide range of direct and indirect assistance to encourage firms to move to disadvantaged areas eg loans, grants, infrastructure improvements.
- European Investment Bank (EIB) provided loans for businesses setting up in disadvantaged areas.
- European Social Fund (ESF) assists with job retraining and relocating.

(b) Evidence might include the marked differences in GNP per capita between the "North" and the "South". Figures should be quoted and regions named, Abruzzi, Molise, Sardinia, Campania and Sicily stand out as being particularly disadvantaged compared to regions such as Emilio Romagna or Lombardy in the "North". Regions such as Umbria and Latium could be said to be in a middle category or transitional. The concentration of industry, commerce and services in the North should be noted.

(c) Answers will be dependent on the country chosen. For Italy, the following factors may be mentioned:

Physical factors
- Relief/geology.
- Climate/water/resources.
- Soil quality/soil erosion.
- Natural disasters.

Human factors
- Remoteness/isolation/communications.
- Limited employment opportunities in the South.
- Decline of traditional industry.
- Land tenure problems.
- Unskilled labour, poorly educated workforce.

(d) (i) Once again answers will depend upon the country chosen.

National Government Measures include:
- Regional development status, Enterprise Zone status, capital allowances, training grants, assistance with labour costs.
- Specific assistance to former coal mining/iron and steel areas.
- Intervention of national government resulting in the relocation of major government employers or state owned firms to disadvantaged areas eg Fiat to Southern Italy, DVLA in Swansea, MOD in Glasgow.
- In Italy the Cassa il Mezzogiorno would be a key policy.
- Tesco Finance to Glasgow - £5 million Regional Selective Assistance (RSA) grant.

(ii) Comment should be made on the effectiveness of the measures outlined eg the long term benefits or disadvantages of using these incentives.

Question 6 – Development and Health

(a) (i) Candidates should be able to identify several differences between provinces using the figures provided. It is clear that the North Eastern province is by far the least developed across the development indicators and that Central province is clearly at the highest level of development within Kenya.

Candidates should get credit for noting that the table covers the three major areas of education, wealth and health and could comment on each of these in turn.

- Education – varies from 87% of females with no education in North Eastern province to only 10% in Nairobi. The striking difference between male and female percentages, especially in poorer provinces, with males getting preferential treatment in many developing countries, could be noted.
- Wealth – all areas of Kenya have many poor, but again big variation from almost 2/3 in North Eastern, Western and Nyanza to <1/3 in Central.
- Health – huge variation again here with >3/4 of children in Central province having all vaccinations whereas only 8% in North Eastern are protected. The North Eastern province trails all others alarmingly in this indicator.

(ii) Answers will, obviously, depend on the Developing World country chosen but for Brazil could include:

- The South East is much more prosperous than other regions due to the concentration of industry and commerce in the "Golden Triangle" of Sao Paulo, Rio de Janeiro and Belo Horizonte. This area has the best transport system in Brazil, the greatest number of services, and has benefited most from Government help. Coffee growing has long been carried out on the rich terra rossa soils around Sao Paulo producing job opportunities and creating wealth for the area and the national economy. Rio de Janeiro – until 1960 the capital of Brazil, had the advantages of a good natural harbour which encouraged trade, immigration, industry, and more recently, tourism.
- The North East, in contrast, is handicapped by more negative factors such as periodic droughts, fewer mineral resources and a shortage of energy supplies all of which have encouraged outwards migration.
- The North (Amazonia) suffers from its more peripheral location, its inhospitable (rainforest) climate, poor soils, dense vegetation and inaccessibility. Not surprisingly, it is the poorest of Brazil's five main regions. Until recently, there was a lack of Government investment and much of the region lost out on basic services such as health, education and electricity.
- In addition to explaining the sorts of marked socio-economic regional variations which exist in a huge and diverse country such as Brazil, candidates may also comment on the marked differences in living standards which exist between relatively wealthy and better-provided-for urban areas compared to poorer more isolated rural areas and to the contrasts that can be found *within* urban areas – eg hillside favelas such as Rocinho in Rio versus the prosperous apartments overlooking Copacabana Beach.

(b) Measures used to combat the spread of malaria can include:

Trying to eradicate the mosquito/mosquito larvae:
- Pesticides/insecticides such as DDT and later Malathion.
- Mustard seeds thrown on water areas become wet and sticky and drag the mosquito larvae under the water, drowning them.
- Egg-white sprayed on water surfaces creates a film which suffocates the larvae by clogging up their breathing tubes.
- BTI bacteria grown in coconuts – the fermented coconuts are broken open after a few days and thrown into the mosquito larvae-infested ponds. The larvae eat the bacteria and have their stomach lining destroyed.
- Putting larvae-eating fish such as the muddy loach into ponds.
- Draining swamps, planting eucalyptus trees which soak up excess moisture, covering standing water.
- Genetic engineering, eg engineering sterile male mosquitoes.

Treating those suffering from malaria:
- Drugs like quinine, chloroquine, larium and malarone.
- Quinghauso extracted from the artemesian plant – a traditional Chinese cure.
- Continued search for a vaccine – not available as yet.
- The WHO 'Roll Back Malaria' campaign.
- Research carried out by the Bill and Melinda Gates Foundation.
- Education programmes in:
  - the use of insect repellents such as Autan
  - covering the skin at dusk and dawn when the mosquitoes are most active
  - sleeping under an insecticide-treated mosquito net
  - mesh coverings over windows/door openings.

(c) Primary Health Care (PHC) strategies may include:
- Use of barefoot doctors ie trusted local people who can carry out treatment for certain common illnesses, often using cheaper, traditional remedies.
- Use of Oral Rehydration Therapy (ORT) to tackle diarrhoea and dehydration which can kill babies and young children.
- Vaccination programmes against diseases such as polio, measles and cholera. PHC can focus on preventative rather than more expensive curative medicine.
- Health education schemes in schools and communities, targeting children and women in relation to hygiene and diet. Using songs, posters, word of mouth rather than written information in societies with high illiteracy, especially among women.
- Local initiatives backed up by small local health centres staffed by doctors who can refer more serious/complex cases to hospitals.
- PHC can also be involved in provision of clean water supply eg WaterAid's work in Tanzania. Also construction of pit latrines/Blair toilets for decent sanitation, often with community participation.

Hey! I've done it

BrightRED
PUBLISHING

© 2011 SQA/Bright Red Publishing Ltd, All Rights Reserved
Published by Bright Red Publishing Ltd, 6 Stafford Street, Edinburgh, EH3 7AU
Tel: 0131 220 5804, Fax: 0131 220 6710, enquiries: sales@brightredpublishing.co.uk,
www.brightredpublishing.co.uk

Official SQA answers to 978-1-84948-215-8
2007–2011